S0-BVP-302

3·21·72

SOUTHERN HORIZONS

WILLIAMS HAYNES

SOUTHERN HORIZONS

Essay Index Reprint Series

BOOKS FOR LIBRARIES PRESS
FREEPORT, NEW YORK

Reprinted 1971 by arrangement with
Van Nostrand Reinhold Co.

INTERNATIONAL STANDARD BOOK NUMBER:
0-8369-2366-9

LIBRARY OF CONGRESS CATALOG CARD NUMBER:
78-152174

PRINTED IN THE UNITED STATES OF AMERICA

Affectionately to

BETSY AND DICK

Contents

A Warning to the Reader

THIS BOOK needed to be written. There is no doubt whatever about that. And yet I wish that I might sit down with you and chat about how it came to be written and by a Damyankee from Connecticut.

A revolution is brewing in Southern farms and factories, and it is high time that all of us look into its causes and consider its effects. A new economic era is beginning in which new crops will be grown and new goods manufactured. More important than the new techniques or the new working conditions or the new markets is the new spirit that today inspires the South. I have attempted to reappraise Southern resources in the terms of these new values.

But you should be warned that this is no exhaustive economic treatise. You will look in vain for charts of prices and wages or statistical tables of production and consumption. This is not that kind of book at all.

This is a story of men, practical men of vision. It is a first-hand account of some of the leaders in this Southern revolution; revealing stories of the men who are doing unusual things in fresh ways. These are the men who are trying to grow cottonless cotton plants; men who are skimming oil off the waste black liquors of the paper mills and using it in soaps and paints; men who are blasting the atoms in petroleum with electrical charges to make undreamed-of products.

To show you what they are thinking, and how they work, and the prizes they hope to win, I have placed each in his workaday world. Hundreds of others who are taking active parts in the great forward thrust of Southern life, I have been forced to omit. And I have had to pass by some developments that may rise up to smite me by becoming important. But by means of this simplification I have hoped to give you a vivid impression of what is happening in the South and a clearer understanding of what it means to the whole nation.

This book is the outgrowth of a long journey collecting material for some articles for *The Saturday Evening Post*. Ten years ago I covered much of the same ground, talked to many of the same men, and wrote a series of articles in which I forecast the coming chemical industrialization of the South. At that time my prophecy was pooh-poohed by some skeptics. Today the writing of the following chapters has been a particularly enjoyable revenge.

Nor have I been merely an inquiring reporter roaming strange territory. I went South to college; an experience I heartily recommend to young New Englanders, and in reverse, to young Southerners. Later for several years I worked intimately with a group of Southern industrialists as their chemical consultant.

A WARNING TO THE READER

The first time I visited the Tennessee Valley, during World War I, Muscle Shoals was a few naked concrete walls surrounded by a sea of mud through which sloughed gangs of mules dragging in construction materials. I saw the two pioneer alkali plants of the Deep South, at Corpus Christi and Lake Charles, when they also were under construction.

Many of my old, close friends are scattered from Newport News to Fort Worth, from Natchez to Tampa, and I have enjoyed Southern hospitality in the old-fashioned meaning of that distinguished but abused phrase, from the pillared mansions of Maryland's Eastern Shore to the sprawling ranch houses of the High Plains of Texas. Over innumerable cokes and juleps we have discussed all sorts of things from the soul of man to insecticides.

One thing above all others that a Southerner resents is the visitor who spends a couple of days in Richmond and Atlanta, a week in New Orleans, a weekend at Colonel Pendleton's plantation, and after driving a couple of hours over Tobacco Road goes home and writes a book about the South with explicit directions as to just what Southerners should do about it. A New Englander can sympathize with this resentment. We, too, have suffered from these "mental carpetbaggers." Our climate, our ancestors, our industries, how we live and what we think, have now been extravagantly praised, now censured beyond all reason.

A WARNING TO THE READER

From Baltimore to Brownsville so many think that quite the damnedest of all Yankees hail from my native Connecticut, that I have become a little self-conscious. Maybe this is only a sort of inverted state pride. Nevertheless, I am so frequently twitted about our blue laws and wooden nutmegs, and the Connecticut Yankee is so universally accepted as the perfect exemplar of the New England conscience which will swap one biscuit for one barrel of flour, even, any day, that it troubles me at times.

This enforced humility has at least kept me from the grievous fault of offering gratuitous advice. Again you must be warned that in the following pages you will find no ready-made plan of economic salvation. You may be surprised that there is so little consideration of the South's social and political problems which always press so closely upon all industrial and agricultural developments. I have tried to keep strictly to the technical and economic phases, within the bounds of my experience.

It would not do to belabor the point, but Southerners and New Englanders stand upon a good deal of common ground. Most importantly we both live in established communities where we have not only known our contemporaries since childhood, but we also remember their grandparents. Our homes, our schools and churches, the bank and the corner drugstore, our highways, even the land itself, are all primed with memories that keep tradition alive. We have an accumulated experience that warns us to welcome the new, be it a gadget, or a man, or an idea, as something to be tested and proved. It is a habit of thought that provokes some of our fellow Americans. But then, it seems to us that they frequently make the careless mistake of identifying change with progress. If the atmosphere of the settled community breeds conservatism, it also engenders faith in America's past and a vibrant confidence in its future.

Southerners are throwing off that deadening apathy which is a pernicious variety of in-growing conservatism gone to seed. Their faith has been reaffirmed; their confidence, rejuvenated. To a New Englander sensitive to this cast of thought, the great revolution in the South is that Southerners are looking, not backward, but forward.

That is why I have called this book "Southern Horizons."

WILLIAMS HAYNES

Stonecrop Farm,
Stonington, Connecticut

1

Revolution in the South

IN ONE of his few inept phrases, Franklin Roosevelt once hailed the South as "the nation's economic problem No. 1." While everybody enjoys hugging the notion that his own problems are peculiar and most difficult, nobody likes to be considered a problem to his family and neighbors. So Southerners bitterly resent this well-meant but tactless statement.

Donald Nelson, Roosevelt's own chairman of the War Production Board, swung to the opposite pole and declared the South to be "the nation's economic opportunity No. 1." In a speech at Atlanta he inventoried the South's wealth of resources and cast up a tidy sum of postwar prosperity. This did not impress Southerners. That particular flattery no longer tickles. Southerners know all about their abundant raw materials, their salubrious climate, the new war plants that dot their landscape—but so what? Today most thinking Southerners are as disillusioned and realistic as a surgeon at the operating table.

Both these distinguished opinions cannot be right.

Obviously the South has enjoyed a great war boom. But in the postwar years ahead of us will this rich region slump back or forge ahead?

Southerners have grave economic problems, plenty of them, and they do not always agree as to either causes or cures. Whether, for example, a low average of family income—lower in the South than in any other important section of the nation—is the root or the fruit of their economic ills is as warmly debatable among them as the classic riddle of hen and egg. It was editor Peter Molyneaux of the *Texas Digest* who laid bare for me the South's real economic dilemma. Ignoring both root and fruit, this steel-brained, honey-hearted breaker of idols from Dallas described the seed in a bright paradox.

"The curse of the South is its blessing of abundant raw materials. Because it has no raw materials New England lives off the rest of the country by its wits: the South lives by charity because it has permitted the rest of the country to exploit its natural resources."

This is the Molyneaux technique: a barbed needle jabbed under the skin to stimulate the brain cells. By means of a picturesque parable this same thought was amplified by another brilliant Southern analyst, Frederick H. McDonald, consulting industrial engineer of Charleston, South Carolina.

"If a damned Yankee up in Connecticut," he said with a twinkle in his eye, "found a gold nugget in his backyard, he'd take it straight to the geology professor at Yale, and if it really were gold, he'd build a high fence around the yard so that the neighbors couldn't see what he was doing. Then he'd go to mining with the garden spade and the ash sifter. With the first money he got he would buy the lots of his two next-door neighbors.

"Just before his vein petered out, he'd sell out to a Wall Street syndicate for a cool million, with which he would buy a little chunk of the stock of his local bank, so as to become a director and know what was what in his home town. The balance he would invest in a good Hartford insurance company, and four snug little Connecticut plants making brass piping, ball bearings, plastic cigarette holders, and wooden nutmegs. About this time he might tell his wife what he was up to."

We grinned cheerfully at each other over this Scottish summary of a New England business career, and McDonald went right on.

"If a Southerner found a gold nugget in his garden, he'd set his yardman, Woodrow, to digging up the magnolia beds while he ran to show his find to black Aunt Sally in the kitchen, and to his wife, and to his three beautiful daughters. Then he'd take it down to the president of the bank. The next week the local paper would report that he and the banker had gone to Richmond or Atlanta or New Orleans about his gold mine.

"They would confer with the second cousin of the

banker's first wife, a gentleman with nice contacts up North, and in the end he'd sell out to some Western mining interests with offices in New York. But by the time the banker and his dead wife's cousin and the Northern contacts were all taken care of, he would net about one hundred shares of the mining company stock and $13,750 in cash with which he would purchase his uncle's old plantation, Belmont Hall. After two poor cotton years he'd have to sacrifice his mining stock to save the mortgage."

I laughed aloud at this neat Southern extravaganza, but McDonald did not even smile. His alert face was grave, almost melancholy. "It is not really funny," he said.

"But, thank God," he added devoutly, "we're learning from the Yankees. We are doing more digging for ourselves. We are going to do more of our own selling in the future."

This is an old hope, oft deferred. Southern resources have long been famous. Soil and climate, timber and minerals, coal, natural gas and petroleum, and water-power sites have been prospected, surveyed, catalogued, and thoroughly publicized numberless times. The South's need of industries was recognized before the Civil War, and I have heard Southerners boast that their State Commissions and Chambers of Commerce have spent more time and money on industrialization campaigns than all the rest of the country.

Twenty years ago I listened enthralled to the gospel of a diversified agriculture preached by Clarence Poe to a great audience of Carolina farmers. Their shouts and whistles pronounced a clamorous "Amen." Yet who in the South would deny that a single cash crop, be it cotton or tobacco, peanuts or oranges, is still a potent, pernicious influence?

Have there been changes in this familiar Southern situation? Is the angry protest that the South is not economic problem No. 1, but our first economic opportunity, anything more convincing than the old claim of the professional boosters? Is there really a new pattern of Southern thinking that justifies Frederick McDonald's gratitude to the Almighty?

During the past dozen years there have certainly been many great changes throughout the South. The outward and visible signs of these have come as benefits from the New Deal or results of the World War II effort. Political largess has been showered upon this section. Throughout recurring emergencies, relief has been more than generous. Millions have been spent in various reclamation and conservation programs. It is a poor Southern town, indeed, that has not a new post office or school. During the war, for reasons admittedly as political as strategic, factories, camps, and airfields brought more new jobs and fresh dollars to the South than to any other region.

The South has received these generous economic and

social gifts, not with docile gratitude, but in a spirit of resentful criticism. Most Southerners, in all sections and from every stratum, reject scornfully the entire New Deal philosophy. They do not hesitate to question openly both the lofty aims and the tangible results of its policies.

From Lubbock, Texas, to Raleigh, North Carolina, I talked with many cotton men—big planters and little sharecroppers; ginners, oil crushers, and compressors; brokers, exporters, and mill men; scientists in laboratories and experiment stations—and without exception they agree that the Government's efforts to help the cotton grower are strangling cotton. All maintain stoutly that every Government move in crop restriction, price maintenance, and export bounty has hurt American cotton either as a fiber in industry or as a commodity in world trade. I found not one among them who openly repudiates the Government cotton program. Most of them admit, "off the record," that they are working actively to secure extensions of the conservation benefits, higher parity price and greater loan values, bigger bonuses for every bale exported. Whatever they believe, they act to get while the getting is good. As for the future, one of the wisest, shrewdest of them answered for all when he shrugged his shoulders and quoted Louis XV, "*après nous le déluge.*"

This flat contradiction of thought by deed is, of course, not peculiar to the South. It is a form of mental dishonesty that has become a virulent disease of the American people and its dangerous symptoms are widespread in every section. In the South, however, it does have peculiar causes and effects. There it becomes another of those glaring Southern paradoxes, an enigma that plagues both friends and critics.

To any Southerner, the very thought of a powerful, centralized, paternal federal Government is instinctively repugnant. In Virginia or Georgia or Texas, you will hear more savage denunciation of the New Deal and all its works than in Vermont or Pennsylvania or Iowa. Nevertheless these same noisy critics continue faithfully to support a political machine geared today to override State rights. They vote for candidates who, to state it conservatively, do not share Thomas Jefferson's convictions of individual liberty or his abiding faith in the common man.

For reasons that appear to them good, Southerners have not bolted the Democratic party. For reasons that they know in their hearts are not good, they have followed its programs and accepted its benefits. This is not a happy, comfortable mental attitude. Its inner stress and strain are painful to many Southern men and women.

We Americans are not self-analytical; Southerners no more than the rest of us. Recently a British friend came to me in great perplexity because of our lack of critical expression. In this country for the first time on an important mission, he was naturally anxious to learn whatever he could about a number of prominent industrialists with whom he must negotiate large contracts.

"It's most astonishing," he complained, "when I ask what sort of man Robert Jones is, everybody tells me he is a 'swell guy.' When I inquire about Henry Smith, everyone says he is a 'louse.' I admit that 'swell guy' and 'louse' are brilliantly descriptive slang, and I have come to identify them as a 'ripping chap' and a 'bounder.' But they give me no idea even whether these men are tall or short, fat or thin, fair or dark. What I should really like to know, of course, is whether they are trustful or suspicious, frank or reserved, proud, keen, gullible. Only by the most painstaking cross-examination can I uncover even these elementary traits. As for anything like an apt characterization—

"I know," he continued, "that Americans are not all either saints or sinners. Must I conclude that you are all color-blind; that you register only black and white?"

Whatever our national tendency toward extreme simplification, Southerners admit readily enough the inconsistency of supporting at the polls a party that no longer represents their convictions. Yet too few recognize the discord of principle and practice that rages within themselves. They invent glib explanations, based upon a variety of expediencies, which however satisfactory as excuses, do not ease the inner tension.

One of Mississippi's Congressmen was taken to task

several years ago by the state's much admired Tax Commissioner, Alfred H. Stone, for having supported the bill to enlarge the membership of the Supreme Court. Stone had been a close friend of the youthful Representative's father and he talked like a Dutch uncle.

"But, Mr. Alf," protested the lawmaker, "I didn't approve of packing the Supreme Court. I only voted for it because I had to. After that bill was introduced Jim Farley called all us Mississippi representatives to his office. He told us that although he thought it was a thoroughly bad business, nevertheless the President had his heart set upon it, and that he had orders to tell us to support it or lose federal patronage. The first thing that will happen, he warned, will be that the C.C.C. comes out of Mississippi. You see how it is. What could I do?"

"Boy, let me ask you just one question. If the likes of Jim Farley put that kind of proposition up to Jefferson Davis, or Governor John Marshall Stone, or Senator Roy Percy, or any of scores of honest Democrats, born and bred in Mississippi, who have represented this state in Washington, what do you think they would have done?"

"Oh, Good Lord! Jim Farley wouldn't dare talk that way to those men."

That shocked answer is a naive confession. Afraid that an open break with the ascendant New Deal Democrats would leave them isolated and politically powerless, Southerners who are in revolt against the New Deal philosophy, barred from direct action, seek other outlets. By means of this natural inversion I am convinced that a profound change is being wrought in the thinking of thousands of Southern men and women, not inspired by the New Deal, but irritated by it. Their fellow countrymen will find that out of their political apprehensions, their uneasy conscience, their bitter resentment, a new spirit is being created.

These are strange—but very strong—ingredients, and the new spirit in the South is at once complex and compelling. It comprises a fresh point of view and an urge to direct action. It reaffirms an ancient faith in the importance of independence and expresses a new belief in the value of self-reliance. It is confident, not boastful. Having agreed that Heaven helps those who help themselves, more and more Southerners have no time for whistling by the graveyard. This is quite the most important change in the South in the past two decades, possibly the greatest change since the bitter days of Reconstruction.

The South fought the War between the States to the utter exhaustion of her physical and financial resources. In 1865 she was physically bankrupt. Spiritually, though beaten, she was unbroken. The faith and courage of Southern men and women which sustained them so magnificently through four years of unequal struggle began gathering together the fragments of assets to re-build a new Southern life and economy. The much maligned sharecropper system was the natural answer to the challenge of the enormous economic and social revolution forced by the freeing of the slaves. The plantation owners had land and some farming tools, but no money to hire labor. The freed Negroes had nothing except their naked hands and training in cultivating the great staple crops of cotton and tobacco upon which the whole Southern economy had been based.

"Stay in the cabins," said the landowners. "Till the fields you have always done. As in the past I will provide the ploughs and hoes, the mules, seed and fertilizer. You must eat, but I can no longer feed you, so I will mortgage the harvest in order that we can all keep body and soul together till I can sell the crop. Then I will pay back what I have borrowed and we will divide what is left over."

It was just as simple as that in the beginning, and just as inevitable—what else could either do?—and just as fair, the very fairest of all possible labor contracts, division upon the basis of profit-sharing. But from the very start it struggled under the load that has sunk more brave sound business enterprises than all other reasons combined: inadequate working capital; the load of debt. If you would understand the sharecropper system, read Chapter XXI of Will Percy's wholly delightful, remarkable *Lanterns on the Levee.* Do not forget, however, that long before his time, this system, already loaded with debt, was thwarted and twisted during the early years by powerful, sinister forces. After the war the carpetbagger and his scalawag allies, white and black, descended upon the South and for a long decade picked the fragmentary assets bare, searing the proud, brave spirit of Southerners with bitterness and hatred, an appalling dread and a terrifying sense of injustice.

During the half century between the surrender at Appomattox and the German violation of Belgium, Southern agriculture gradually recovered. But a debt-ridden sharecropper system was a poor basis upon which to establish a permanently flourishing crop economy. Exploitation of land and labor was always a terrible temptation; often a dire necessity. The land was mined and living conditions sank. Cotton followed the Southwestward migration that had begun long before 1860, and while fertilizers helped stave off the evil day of soil exhaustion, the end was eventually inevitable. The foundation of the greatest agricultural area of the country, the land itself, was washing away.

The other natural resources of the South: the pine forests; the coal and iron of the Birmingham district; the natural gas of West Virginia, then Louisiana, finally of Texas; the petroleum of Texas and Oklahoma; were all developed by Northern capital. Not always, but far too often, they were relentlessly exploited. No doubt absentee ownership encouraged the thoughtless, selfish, get-rich-quick habits of those days; but the significant aspect of this highly profitable development of Southern resources was that although it brought money into the South, it was paid out for vanishing assets and current labor. The profits which might have been available for reinvestment were drawn away to other sections.

During these years there was also a growing industrial development. Wise Southerners had long realized that an industrial supplement to their agricultural economy was essential, if the region's admittedly low scale of living was to be raised. The fiery little Celtic-Georgian, Henry W. Grady, was only the most effective of many preachers of this industrial doctrine long before the New England cotton mills, tempted by proximity to their raw material and low wages without unionism, began to trek into the Piedmont sections of the Carolinas where water power was available.

Still lacking capital, but convinced of the need of bringing any and all kinds of manufacturing enterprises into their midst, Southern communities outdid the most bumptious, hustling Western city in ballyhooing their "natural advantages" and in raising slush funds to bribe industrial prospects with landsites and buildings and free taxes. Florida, not California, was the first state in the union to vote state funds for an advertising appropriation. Other Southern states followed and during the booming twenties, when this industrial proselyting reached a giddy climax, all had Industrial Development Commissions of some sort, actively backing an army of local Chambers of Commerce. Though sometimes more enthusiastic than purposeful, this municipal boosting did bring results. If sometimes the new factories were economically wobbly, this generous, high-spirited movement furthered the industrialization ideal.

The Great Depression withered these industrial efforts and blighted the agricultural economy which was still the mainstay of most of the people. The average farm-family income—never more than a couple of hundred hard cash dollars a year in the South—dropped to ninety-five. To avert stark starvation the federal Government had to step in with relief programs which during the Roosevelt administration grew and multiplied and diversified till they came to dominate the work and the thought of great segments of the people. Consciously or unconsciously, for good or evil, the Government not only provided the mainspring of an economic revolution in the South, but it has wound this mainspring up so tightly that it now exerts terrific force.

Wholly intangible, and thus unlike other products of the New Deal—gigantic dams athwart the Tennessee Valley; big, yellow brick schools lording it over so many small towns—this change is not without its own signs and portents. The political revolt that dethroned Talmadge in Georgia and the establishment of the Southern Research Institute are both symptoms of this revolution. A most auspicious omen is an entirely new attitude towards raw materials. Nothing will have a greater influence upon the future economic development of the South.

Conservation of Southern natural resources has long been a worthy cause faithfully supported by a few enthusiasts for the common good. They have not had much active sympathy from their neighbors who have generally regarded them as mildly demented, wildly agitated visionaries. They have had little help from the men and corporations who control these valuable assets. There have been plenty of instances of typically American exploitations of raw materials, but by and large, the attitude has been pretty much that of the wife of a Georgia cracker when approached by the forester of the nearby paper mill. She listened attentively to his plea to prevent forest fires and his explanation of the benefits that would accrue to them if the unburned land were allowed to reseed itself. "Why, in seven or eight years we'll be coming around and paying you a pretty penny for pulpwood. So if there's a fire in your woods, call on us. We'll come over and help you."

"We don't need no help from you, Mister," she replied with spirit. "Doan worry about us. My ol' man'll get all our land, an' plenty of yours, burned over before the spring rains. He reckons to feed 'bout twenty more head of cattle this year and he shore do need the grass."

Put thus baldly, to burn the woods for the sake of the grass seems ridiculously like Charles Lamb's famous Chinaman who burned the stable to get the roast pig. Yet that paper mill forester is not laboring in vain. The mere fact that he is on his job is significant. His employers plainly have a very different conception of timberland from that held by their predecessors, the lumber companies whose motto was "Cut out and get out."

It is one thing to regard a pine forest as a rich bonanza to be gathered as quickly and cheaply as possible and squandered recklessly. It is a very different matter to consider it a bank account that must be built up carefully with an eye to future investments. A small storekeeper in Mobile underscores all that this switch in viewpoint means. He is no "lumber baron." I doubt if he is even a shareholder in one of the large paper corporations. He is just a "little guy," a Southerner of undistinguished ancestry and modest circumstances, one of a great group that is almost unknown but which must not be ignored, a man who has felt the stirring of this new spirit. In an old seaport historically famous for its exports of lumber, rosin, and turpentine, he has caught the new vision of Southern resources and woven it into his own dreams for the future.

"Near a little farm upcountry that my wife inherited from her aunt, I'm buying up all the old cutover timberland I can afford to," he confessed. "Every six or eight years I have a crop of pulpwood; in twenty or twenty-five, some good saw lumber; any year I can rent the gum turpentine crop. The paper mills or the lumber dealers will come in and cut and cart the wood I want to sell them. All I need is a tractor to keep the fire lanes open and some axes to keep the trash growth down. And when I get a thousand acres I reckon I can live out the rest of my life on that land. I'll have troubles enough to keep me thin, but I won't have to go on relief to keep from starving. Looks to me like pine trees is the closest thing on earth to a perpetual cash inventory."

This switch from bonanza to bank account is even more impressive in another area of conspicuous exploitation of Southern resources. Fifteen years ago Louisiana and Texas took over control of crude petroleum and began to prorate the withdrawal of oil from various fields. The petroleum companies had to take it. They have come to like it.

Texans and Louisianans watched that proration battle from the sidelines. The real fight was fought in the state legislature. Many newspapers crusaded actively for the conservation laws, and demagogic talk about "big Northern corporations" and "irreplaceable reserves" and "God's gift to the people" raised the semblance of a popular, political issue. Nevertheless, the people remained blithely indifferent to the oil conservation issue, and their interest in natural gas seldom went further than the rate charged them by the local gas company for its consumption in their kitchen range.

Today, however, you can meet men on almost any street from Brownsville to Jacksonville who feel that they have a financial stake in the oil and gas resources of the Gulf Coast region. Conservation has become a personal dollars-and-cents issue. The altruistic appeal that the waste of these valuable reserves ought to be prevented has become an active demand that these natural raw materials be kept within state boundaries and developed for the benefit of the homefolks.

Just listen to a former Governor of Louisiana, talking in Houston:

"You people in Texas are to suffer the greatest blow. Your gas fields will become the happy hunting ground of the Eastern industrialists. There is pending an application by the Reserve Pipeline Company to take your South Texas gas. There is pending an application by Hope Natural to take your North Texas gas. There is pending a pipeline to California to take your West Texas gas. The Big Inch and the Little Inch oil pipelines would be converted to natural gas to deplete and exhaust your East Texas fields. United Power & Light Company would dip into the fields of the Southwest to supply Iowa, Wisconsin, and Michigan with their natural gas. The Republic of Mexico wants more of your gas and there is willing American capital to pipe it down.

"Your doom will be sealed in a matter of months, instead of years. Because it takes only months to barter away your natural gas forever. Once the pipelines are laid and the contracts are made, your reserves are beyond your grasp, and your industrial future is bartered away for a mess of pottage."

And Sam H. Jones went on to "document," as he himself said, the industrial future of the Gulf Coast by pointing out that natural gas is the key to the future of synthetic rubber and 100-octane gasoline, that it is an industrial necessity as a fuel in a region lacking coal and water power; and most important of all, that it is the coming raw material for the modern plastics and chemical industries.

"As a basic material for chemicals, natural gas simply has no competitor. This alone should build hundreds of industrial plants in gas-producing states. This alone will provide hundreds of thousands of jobs. It will bring a caravan of prosperity. One basic plant will beget many subsidiary plants. It is our nearest approach to a balanced economy–" a lawyer and a politician, expounding to a crowd of Texas businessmen the doctrine of chemical values.

Twenty years ago, an illustrious Georgian, aptly described as "a Southern gentleman who happened to be a damned good chemist," expounded that same doctrine and was laughed at for his pains. Dr. Charles H. Herty deserved a hearing in the South. He had been professor at the University of North Carolina. His improved cup for collecting turpentine made many extra dollars for thousands of Southerners. From his research at Savannah in pulping pinewood came the great Southern paper industry. Yet no one in the South believed him when he warned that with purified cellulose from spruce wood selling at five and one-half cents a pound, the price of cotton cannot be much higher. Cotton and wood, this chemist pointed out, are both essentially cellulose, and cotton planters must meet the competition of wood. His was then a voice crying in the wilderness.

Since Herty made that prophecy, American production of cotton has been cut in half, and still the surplus bales pile up. Domestic output of wood rayon has multiplied by five, but the demand is still insatiable. Then cotton sold for thirteen cents a pound; today the price is pegged at twenty-two cents. Rayon staple fiber then was priced at sixty cents; now it sells for twenty-four cents.

Southern politicians and Washington bureaucrats have still not the faintest, glimmering idea of what such chemical values mean. But cotton growers know now, and there are thousands of Southerners who comprehend exactly what Dr. Eugene Schoch, veteran chemistry professor at the University of Texas, means, when he says bluntly: "If we add five cents per one thousand cubic feet to the cost of natural gas at the well mouth—pouf! we snuff out all its uneconomic uses as an industrial and household fuel, and make it what it really is, an invaluable raw material for gasoline, plastics, medicines, and scores of chemicals."

We stand on the threshold of a chemical age. Almost every Sunday the magazine section of any newspaper tells thrilling tales of new triumphs won in the chemical laboratories. The chemist—to his infinite amusement—is hailed as the modern miracle worker. Wonder products of synthetic chemistry are promised for our clothing, our home, our car, our gardens, our medicine chest. Discounting all this popular ballyhoo, it is a fact that the brightest horizon of our industrial future lies at the end of the chemical path. That path leads straight Southward. No other section of the country is so blessed as the South with all the chemical raw materials. Southerners are learning this, and they are reappraising Southern natural resources in terms of chemical values. They are determined to end the curse that burdens Peter Molyneaux.

2

Competitors of Cotton

GREENWOOD, MISSISSIPPI, is the headquarters of W. M. Garrard, a quiet, dapper man whose knowledge of the cotton business is rather awe-inspiring. He simply must be good. He is manager of one of the oldest, largest farm cooperatives in the United States, the Staple Cotton Cooperative Association. He works for his friends and neighbors in the famous Delta country, the biggest planters in the richest cotton area. They pay him $54,000 a year to sell their crop.

Whether they like it or not, Garrard does not hesitate to tell his members the truth as he sees it. Mostly they like it, because generally he is right. In the autumn of 1944 he laid before them three unpalatable propositions:

First, in spite of all war demands, there is an enormous surplus of cotton: some twelve million bales, a full year's supply.

Second, at six cents above the world market, the price of American cotton has reached an unsound height that cannot be indefinitely maintained, even by the United States Government.

Third, Government aid has heretofore been aimed to help the grower whose position is now better than it has ever been in history, and now it is high time that something was done for cotton.

Garrard's three simple statements make a fine introduction to any consideration of the cotton problem, and from any point of view—historical, economic, or social—cotton is the proper starting point for any consideration of Southern raw materials. Factually Garrard's propositions are correct. Essentially they are true, no matter how sharply the experts may disagree over their meaning or their solution. Such a foundation of fact is desperately needed because pernicious combinations of sentiment and selfishness in the South and of prejudice and ignorance in other sections, have obscured the real and very high stakes that all the American people have in this greatest of all our crops. Sentiment and prejudice are difficult to deal with, but both selfishness and ignorance may be at least enlightened.

Cotton is in a bad way. Most Americans know this and most of us appreciate that cotton is at once our most valuable textile fiber and the most important crop of our Southern states. But few people who live North, East, and West have the cloudiest realization of what all this means in the South.

This thought was driven home at a delightful Sunday

supper party. Our host was Carl Fritsche, the emphatic engineer who operates the Reichhold chemical plant at Tuscaloosa. Fritsche has been a husky trail blazer in the chemurgic movement to use more farm products as factory raw materials and he laid before us his clean-cut opinions about cotton economics. The sociological aspects were firmly, gracefully set forth by Raymond Paty, President of Alabama University, a gentleman of the New School of the Old South. The feminine point of view—important, indeed, when we discuss the uses of cotton—was expounded by our wives with backgrounds ranging from New York to New Orleans.

Into the smooth, easy flow of this conversation Hudson Strode flung a verbal brickbat. Strode has lived abroad and has written thought-spurring books about Finland and Mexico, but he knows Alabama.

"Let us face the facts," he said. "The North not only beat the South in the War between the States, but it ruined the South during Reconstruction and for years since has robbed the South by high tariffs and preferential freight rates. It is only poetic justice if the taxpayers of Massachusetts, New York, and Pennsylvania must now pay us to grow cotton, as we have paid to support their industries."

Fearing for Mrs. Fritsche's nice party I tried to cudgel up some diverting query. Our host himself saved the pleasant discussion by a burst of mock heroics. I caught his tongue in his cheek and relaxed to enjoy the capital show he was putting on.

"We are all forgetting," Carl Fritsche explained passionately, "that cotton is the miracle plant, the only plant that gives man food and raiment, feed for his cattle, fertilizer for his fields. Cotton is God's great gift to the South, given us as a sacred trust for mankind. Dare we repudiate Jehovah himself by refusing to grow cotton for all the world even if our best fields are gullied by erosion and our rich red clay impoverished by overcropping?" He pounded the table and glared at us defiantly.

"Believe me," he concluded with a fine flourish, "we shall continue to grow cotton—at New Deal expense—come the Republican party or a Red Revolution, until death do us part and hell freezes over!"

We all burst out laughing. Someone suggested he should be an actor, and he protested, "Oh, no, not that," adding modestly, "But why not a Congressman from Mississippi?"

Turning to me, he spoke very seriously. "Southern politicians will hand you just such chunks of raw emotionalism, but sober cotton men will also give you this sentimental tripe in place of honest argument."

And some of them did. During the next three months this tirade, stripped of Fritsche's flaming grandiloquence, was repeated several times. Always I remembered the sane comment Dr. Paty made to his warning.

"When sentiment replaces reason it is rooted deep in human emotion or individual experience. In the South, where for generations cotton has been the economic and social base of whole communities, it is as hard for a man to consider cotton objectively as it is to regard his own wife dispassionately."

To many Southerners cotton has long been not only the means of life, but their very way of life. To big planter and little sharecropper, to Mr. Page who runs the bank, to old Dr. Reynolds and young Sam Tolliver, the haberdasher, to the white boy who clerks in the freight office and the black boy who sweeps out the Elite Barber Shop, to the entire neighborhood, cotton was supreme, at once the staple crop and a great historical tradition.

This overemphasis in the Cotton Belt is counterpoised by woeful inappreciation in other parts of the country. Solution of the cotton problem is certainly not helped by the fact that most of our citizens are ignorant of, and so quite indifferent to, the meaning of cotton in the economy of the nation and in the workaday life of each of us.

How many Americans, for example, realize that we use four times as much cotton as all the wool, silk, and linen, rayon, Celanese, and nylon added together? The palpable differences between a gauze bandage and a canvas awning display the versatility of this most used, most useful textile fiber. But until we stop and think of it, how very few of us remember that cotton gives us men's broadcloth shirts and women's gingham dresses; denim overalls and khaki uniforms; sheets, bath towels, and handkerchiefs; sails for ships; belts to drive machinery.

Every American schoolchild learns that cotton is the chief cash crop of our Southern states. But how many of our adults realize that this white fluffiness is grown on fields roughly equaling the total land areas of Belgium, Holland, and Denmark; or that it is the direct support of thirteen million Americans, one in ten of all our people? That is, more human beings than work in any single agricultural or industrial group in the country.

This great crop is in truth two crops: the fiber and our most important oilseed. Thanks to chemical research, this seed has become as versatile as the lint. The Texas Cottonseed Crushers' Association has a traveling display of one hundred and fifty-one products made from what was once a troublesome waste that piled up outside the cotton gins.

Largely because the cotton plant produces this 2-in-1 crop, it returns to the grower more dollars per acre than do any of our great field crops. The average for 1932-41 was cotton, $27.97; corn, $13.79; wheat, $10.32 per acre. That forgotten fact bears remembering.

It raises the provocative question: If cotton yields almost twice as many dollars per acre as corn, why is the average annual income of the farm family in the Cotton Belt only a third of what it is in the Corn Belt? The answer is disarmingly simple. Cotton acres support four times as many people as corn acres. This ignored fact should also be remembered.

It makes quite clear that low farm-family income is not the cause, but the result, of the cotton problem. Confusion of this cause and result led the Government to assume control of the price of cotton for the benefit of the planter, a policy that has transformed what was originally a fairly simple, straightforward relief problem into an international diplomatic issue.

In the good old days when American cotton dominated world markets, we produced some fifteen million bales, half of which were sold abroad. In the late twenties when the world went into an economic tailspin, the cotton market collapsed; demand dwindled; our exports all but vanished; the price fell. Since 1929 our exports have been dropping, dropping in 1941 to a low point of 1,100,000 bales, which was but a tenth of that year's very small crop. In the South this shrinking cotton market precipitated a human crisis.

Paying no rent and growing most of its food, a family on a farm lives in clover on much less hard money than a city family. But in 1930 when family cash incomes in the South shrank to a pitiful $95 a year, thousands who grew only cotton for a cash crop faced dire distress. So following patterns designed for corn and hogs, the Government decided "to do something for cotton." In Henry Wallace's "plough under" theory, this meant a cut in cotton acreage. A short cotton crop would mean a high price. Automatically, this would raise the low farm-family income in the cotton states.

That was how it was planned, but what actually happened was that every grower planted his best acres, bought more fertilizer, cultivated harder. Though four million fewer acres were planted, the yield per acre jumped from one hundred and fifty-seven to two hundred pounds. The crop increased three million bales; the price slithered from nine and one-half cents to four and three-fourths cents per pound. Instead of putting the cotton grower out in the clear, the scarcity program dropped cotton plumb to the bottom of the well.

Since 1929 scores of schemes have been tried to sustain the price of cotton: all sorts of production-control programs by the Agricultural Adjustment Agency, and sundry loans, purchases, and export bonuses by the Commodity Credit Corporation. The price has been shoved up to twenty-two cents a pound. The net result has been to jockey American cotton into a position where it must now fight a three-front war.

World War II imposed a truce in cotton's war, but postwar American cotton must again battle with the low-cost foreign growths in world markets. In the home market cotton must again compete with rayon, Celanese, and nylon in fabrics and with paper in bags and cartons. But the most desperate, most doubtful battle will be on the political front. Here greed and ambition enlist Quislings perfectly ready to betray this great American crop for their own selfish ends.

In the worldwide market, our foreign cotton competitors are sitting snug and happy under the umbrella of our artificially high price. Consistently they cut six cents or seven cents under the American "parity price." Except on lend-lease deals, they naturally get the business. For them it is extremely profitable business, and nobody dares guess how cheaply they could sell if it came to a knockdown, drag-out price battle. Every year, while our cotton acreage shrinks, theirs grows. Under the protection of our own subsidies to American growers, they have raised their production from nine million to fifteen million bales a year. Some nations have made stunning gains. Mexico has trebled her cotton acreage and Argentina quadrupled hers; Brazil has multiplied by six and Russia by ten.

To meet this situation the Commodity Credit Corporation recently adopted an export bonus of four cents a pound. What happened is prophetic—and disconcerting. Brazil immediately cut her export price four cents and protested violently to our State Department at this flagrant, official violation of the principles of the Atlantic Charter calling for free trade in the world's staple commodities.

This export-bonus idea is packed with TNT. It puts

us in a ridiculously unsound position. We are officially committed to global free trade in all the world's great commodities. One of our most persistent complaints to the totalitarian powers has been against their export subsidies. If we do not heed the same protest of our good friends—Brazil today, but it may be Russia or China or Mexico next year—they might humanly retaliate by dumping cotton here at a price far below "parity." Our Government will then be on a very hot diplomatic spot and the domestic backfire will be loud and explosive.

The most cheerful view of the foreign cotton situation is that an expanded postwar international trade will raise living standards in low-income nations so that there will not be enough cotton to go around. Statistically the picture is perfect. Prewar, we each consumed twenty-eight pounds of cotton every year: abroad average consumption was less than five pounds per person. If United States consumption remains at its prewar level and foreign consumption can be raised to, say, fifteen pounds, there is not enough land on earth climatically adapted to the crop to produce sufficient cotton to fill the demand. Even the most optimistic global planner bogs down on such a program to solve the American cotton problem.

Already we have made one tangible contribution to this international program. On invitation from our State Department and at the expense of the American taxpayer, a picked group of Chinese cotton men, students and agricultural experts, have toured our Cotton Belt. They were shown everything we have; taught everything we know. They were given generous samples of seed of our best varieties, including special strains developed to meet special conditions of soil, rainfall, and temperature. It is rumored that under lend-lease they got cultivating machinery, even the latest mechanical pickers, that American planters cannot buy. I happened to follow this delegation through Texas and the comments I heard would curl a pigtail into a pretzel. To Texas growers the mildest, most printable description of this altruistic junket was "a goddam booby trap."

On the substitute's front of cotton's war, the battlefield with synthetic fibers is quite different, but the key to the campaign is the same. From 1920 to Pearl Harbor, American consumption of cotton doubled, but that of rayon increased fivefold. In terms of cotton, this means that domestic rayon production increased from 300,000 to 1,400,000 bales. In terms of dollars, fifteen years ago cotton sold at thirteen cents; today it sells at twenty-two cents. Rayon staple fiber, which then sold at sixty cents, now sells for twenty-four cents. Cotton's boast of being the world's cheapest textile fiber is challenged. Staple rayon is ready for the machines. Cotton in the bale must be cleaned and processed. And after the war, the price of rayon promises—or threatens—to be fifteen cents or lower.

Nor is that the whole story. In 1930, a third of the cellulose used in making rayon, transparent wrapping paper, photographic film, and cellulose lacquers came from cotton linters, the short fuzz of lint that clings to the cottonseed after it passes through the gin. Today, ninety per cent of the cellulose for these purposes comes from wood. In time of war there is an imperative demand for cotton cellulose to be nitrated to smokeless powder. Therefore the War Production Board allotted "chemical cotton" so niggardly that makers of rayon and film and lacquers were forced willy-nilly to adopt "wood cellulose."

"They have clubbed us out of using linters," said Arthur Petersen, who runs the brand-new Celanese chemical plant near Kingsville, Texas. "Now that we are on the new basis, they will jolly well have to coax us back again."

This compulsory war experience has resulted in better, cheaper wood cellulose and a rapid accumulation of know-how in handling this material in the manufacture of fibers, films, and coatings. For the future of cotton it is important, not that wood will be a more serious competitor of linters, but that wood cellulose, the raw material of cotton's direct competitors, will be cheaper and its manipulation more expert.

On cotton's third front, the war of liberation from political control, the fighting has hardly begun. Today the collaborationists are very much in control; tomorrow the politicians and bureaucrats will undoubtedly defend their entrenched positions with all the courage and resourcefulness of a suicide squad.

For ten years now every move in the Government's cotton program has made the plight of cotton more desperate. Cotton producers have been greatly helped, but the product, cotton, has been dangerously hurt. On both the other fronts, foreign and synthetic, it is extremely significant that the artificially high price plays right into the hands of cotton's competitors. Every cotton man I talked with, from North Carolina to West Texas, admits that the only sound, permanent solution of their problem is to get cotton out of politics.

The politicians play the game the Southern farmers are either too greedy or afraid to repudiate. Southern men in Congress are intelligent enough to know that high prices harm cotton, but they are smart enough to know what their constituents want.

At the very outset let us count the blessings of the Government cotton program. They are tangible and important. They have been costly, but surely no good American begrudges any expense to help a tenth of our fellow citizens achieve a more representative standard of living. Ten years of relief measures have lifted the burden of debt from many poor families and helped them escape from the treadmill of the single cash crop. Various conservation campaigns have taught better agricultural practices, helping thousands to achieve greater economic independence. Extensive soil-improvement and erosion projects are restoring and preserving our most vital national resource, the land.

The whole Cotton Belt is becoming more self-sustaining. The acres least suited to cotton are being planted to food and feed crops. In the Mississippi Delta, for example, in 1938 before the control and soil-conservation programs became effective, seventy-two per cent of all tilled land was in cotton: today this is down to forty-one per cent. On this reduced acreage as much cotton is grown as formerly while there is also an enormous production of oats and sizable crops of barley, rye, lespedeza, and alfalfa which support a fast-growing livestock industry. Over in southern Mississippi and Alabama so many white-faced Herefords and black Angus cattle are grazing that one might think he were in Texas. This is all a great gain, but cotton itself, both as crop and commodity, has suffered great losses.

Thoughtful cotton men realize that the Government cannot rescue King Cotton. They know that the best planters on the best acres are "cleaning up on twenty-two cent cotton." They realize that if they employ full modern mechanization—tractors to plow, plant, and cultivate—chemical defoliation, and mechanical picking, they could "make" cotton so cheaply that they need not fear any competition, synthetic or foreign.

In their hearts they know, too, that because these modern methods cannot be applied to the small, gullied fields of the Carolinas, Georgia, Alabama, and southern Mississippi, cotton is simply out of the picture east of the Alleghenies. Old Black Joe with his mule and his hoe and his chilluns picking snowballs in the two-acre cotton patch can no more compete with the tractor moving back and forth over a hundred level acres in the Mississippi Delta or West Texas than a wheelbarrow can compete with a motor truck.

But such cold, competitive facts do not jibe with bureaucratic ideas of farm relief; and the taffy of Government relief is very sweet and habit-forming. Accordingly we have built up in the South a political vested interest in Government interference and coddling which it will be as difficult to eradicate as Nazi ideas from German youth.

From first to last the Government cotton program displays all the symptoms of the cancerous political disease that today threatens every democratic people. Its causes, its methods, the ends it sought and the results it has achieved, if we understand them, it will help us diagnose the disease before it is too late to save the patient. Every move of the Government cotton program has tended to keep poor land and incompetent farmers in production at the expense of the whole American people. We can well afford to be generously charitable. Unless we choose to commit national suicide we dare not be deliberately inefficient. Human values cannot be overappraised, but economic values must be reckoned dispassionately.

Throughout the South one hears savage denunciation of the Government for its failure to solve the cotton problem. Yet it is far beside the point to blame either Southern politicians or Washington bureaucrats for refusing to lay the ax to the tangled roots of the problem of low farm income, or for grabbing the price of cotton as a lever to lift farm incomes without cutting those roots. The effort has failed. It was predestined to failure. That failure should be a lesson to us all, for its causes were political, and in a democracy the correction of political mistakes, like the cure of political evils, rests with all the people.

3

Self-help for Cotton

IN TEXAS a man is growing cotton plants that produce just as little cotton as possible. Practical cotton planters say that Dave Killough is crazy—not as crazy as a loon, but crazy like a fox.

For excellent reasons it makes sense to grow cotton that produces very little white, fluffy fiber but as much of the fat little seeds as possible. Because we have been selling less and less cotton abroad, the surplus bales have piled up higher and higher year after year. Because the demand for edible oil and cattle-feed meal has been steadily growing, the supply of cottonseed has been chronically less than could be sold.

As the planter trucks it to the gin, cotton returns to him for every five hundred-pound bale of lint, one thousand pounds of seed. The market situation being what it is, if a cotton plant might be developed that yielded, say two hundred and fifty pounds of lint and two thousand pounds of seed, lint output would be halved; seed production doubled; and the thirteen million Americans who depend directly on the cotton crop for their living would obviously be better off.

That is why the authorities at College Station, Texas, where Killough is in charge of plant-breeding investigations, do not fire him as a crackpot. Rather, they are encouraging him to develop varieties that produce less lint, but bigger and better seeds. It is a drastic biological effort to help cotton after so many political helps have failed.

At least the idea makes sense, and the big, rawboned Texan who is trying to make it work believes that it can be done. If it seems to you a forlorn hope, recall that every practical cotton man—big planter, little sharecropper, ginner, crusher, exporter—agrees that the Government's cotton programs have failed dismally. In desperation, after fifteen years of "let Uncle Sam do it," he is turning to some pretty strenuous self-help methods.

Plant-breeder Killough has been belaboring his crazy notion since 1940. Like all good scientists he is ready enough to tell what he has actually accomplished to date. He is as shy as a wood thrush in talking about what can still be done and what the outcome will be. He has found—"isolated" is the correct term—six cotton plants that if not strictly lintless can be honestly described as semilint. Several of these produce more, larger, heavier seed than the average. All of these selected specimens reproduce these characteristics.

This cautious Texan—a contradiction rivaling the cottonless cotton plant—suspects that there may be some mysterious correlation between sparsity of fiber and superfluity of seed, but as he prudently puts it, "at least the cotton plant is willing to cooperate, and that's essential."

In one way these six picked plants are very inconsiderate of the problem they are being asked to help solve. They have a bad trick of scattering their seed. When the bolls burst open, the seeds, with so little lint to hold them, are catapulted broadcast. Killough believes this can be corrected, since some plants are less addicted than others to this spendthrift habit. Further progress will require more of the same scrupulous methods of selection and then propagation.

After four years of that eyestraining, backbreaking selection, Killough collected a bushel or more of seed of each of these six varieties. In the spring of 1945 he was able to plant real field tests with the prospect of harvesting sufficient seed cotton for practical commercial trials. Next winter he expects to begin to know a little about what might be expected of his semilint cotton.

Some of his neighbors are not so circumspect. They are sure that cottonless cotton is not a white blackbird, and they see cotton grown for seed as the salvation of the Texas planters. They have always grown the lower grades, with a comparatively low yield per acre, on a large scale at a comparatively low cost. Most of the Texas crop was always sold for export. Killough's bright idea fits the Texas conditions like a bulb in its socket.

One of the most stubborn, troublesome facts underlying the whole cotton problem is that conditions in different sections of the cotton-growing areas are as different as eggs and eggplant. So sharp are these differences that any blanket program, whether for cultivation, marketing, or relief, is foredoomed to failure. Any plan or any conclusion should always recognize that the Cotton Kingdom, like all Gaul, might well be divided into at least three parts. There is a fourth, like the German barbarians across the Rhine in *Caesar's Commentaries,* looming up in the newly opened, irrigated sections of Arizona, New Mexico, and southern California. Historically the three important, distinctive sections are the Southeastern, comprising the Carolinas, Georgia, Alabama, and southern Mississippi; the Delta Country, stretching along both sides of the Mississippi River from Natchez to Vicksburg; and Texas, which, since cotton is fading out of the Gulf Coast section, means today West Texas.

The Southeastern area is the traditional "land of cotton" of song and story. To most Americans it represents the typical cotton country. A realistic insight into its conditions is badly blurred either by the rosy tints of the romantic school or the muddy daubs of the modernists. Neither "Dixie" nor "Tobacco Road" presents a true picture, although both color heavily a lot of our thinking about the whole Cotton Belt, its people, its crops, and the problems of both.

This Southeastern section labors today under exceptional economic and agricultural difficulties. Much of its soil is outworn and it requires heavy applications of fertilizer to raise more than half a bale to the acre. It is a land of small fields, making mechanical cultivation difficult, if not impossible. It has a big population of white and colored tenants, mostly poor sharecroppers who find it exceedingly difficult to raise substitute crops. Here is the tough taproot of the cotton problem, the region where costs are so hard to reduce that in anything like free competition with the Delta and West Texas it must be forced out of the crop.

Nevertheless, it is upon the needs of this most needy section that the whole Government cotton program has been predicated. As a human problem of relief this is right; but it is wrong as an attempted solution of one of the nation's grave economic problems. Government measures have all tended to keep poor land and incompetent farmers in production, and this is as true of corn or wheat or hogs as it is of cotton in every state of the Cotton Kingdom.

Compared with the Southeastern section, the Delta is lush with agricultural opportunity. Level fields of rich soil as black as licorice stretch to the horizon, dotted with the tiny shacks of sharecroppers whose fate delays full mechanization. This section is American headquarters of the premium long-staple cotton whose individual fibers measure more than an inch in length. With all their manifold advantages of soil and climate, of big operations and experience, Delta cotton planters can meet any competition. Besides, they have three aces in the hole.

Except for their labor problem, which to them means an excess, they can thoroughly mechanize their crop. Though there is a labor shortage now, all are confident that the Negroes, lately in war plants, will return to the plantations the moment jobs become scarce.

Second, they can expand operations. Thousands of acres of good, black land are still drowned in swamps. It is naturally the roughest, most expensive land to clear, but I heard of several planters who are quietly acquiring less desirable land and planning postwar to use Army equipment that now builds an airstrip in the jungle in a few days. It will be a simple job of "ridge and slough," digging irrigation ditches to drain the fertile muckland.

Other longheaded Delta men are putting cotton profits into tractors and reapers or blooded beef stock, because if forced out of cotton, their plenteous soil could grow many other crops or fatten cattle. Winter oats and alfalfa are already there. Oats, practically untended, harvest fifty bushels to the acre; intensively cultivated, one hundred bushels. These are figures that give the Northern farmer pause. Even carrying a big Negro population on their shoulders, these Delta planters face the future confidently.

Out on the High Plains of Texas an entirely different breed of men are tilling an entirely different soil. There, as in the Delta, land stretches to the horizon. You drive a mile, two miles, through one man's property. But those long, level fields are not spotted with sharecropper cabins. The Plains have their labor problem, but it has always been one of shortage, hence they began to mechanize years ago. Today, except for harvesting, they are thoroughly mechanized upon a four-rows-at-once cultivation system. The mechanical picker is here, too, three good working models from the implement houses, and scores of budding geniuses are working nights and Sundays in barns and garages, tinkering upon an idea worth millions to the luckiest inventor.

Texas cannot, and does not try to, grow long-staple cotton. Theirs is all mass production, a conception so deeply impressed on them that they are the only large group of agriculturists I have met anywhere who habitually speak in terms, not of bales or bushels or tons to the acre, but of men to the acre. They produce middling grades of cotton—most of which was previously exported—made with a sharp eye on cost. They are cost-minded rather than price-minded, and recently two experts of their State Experiment Station, C. A. Bonnen and A. G. Magee, published an exhaustive study of growing costs in various sections of Texas. Out on the High Plains, the low-cost area, they calculate that on a thoroughly mechanized farm, including machine picking, with two-row, tractor-drawn equipment, cotton lint can be made at a cost of four cents a pound; with four-row equipment, three and eight-tenths cents.

With cotton today priced at twenty-two cents, those are pretty shocking figures: shocking in one way to the housewife, who, shopping for hubby's handkerchiefs, finds it almost impossible to buy for fifty cents a poorer grade than she formerly bought for a quarter; shocking in quite a different way to the cotton grower in other sections where costs are twice, even three times, as high. Yet in Texas nobody seriously questions these figures, and I met a number of growers who admitted, "I'd hate to do it, but if I had to, I guess I could sell cotton for six cents."

It is easy to see why Texas planters think that Dave Killough is crazy like a fox and why they are so eagerly waiting the outcome of his experiments with his six selected strains of semilint cotton. There is really more to his idea than less cotton and more cottonseed from the same acreage. A semilint cotton would by-pass the entire ginning operation and go straight to the crushing plant. Here its leaves and burrs would be removed in cyclone separators. The scanty fiber could be recovered in the regular delinting machines. The operations of dehulling, cooking, and pressing out the oil might be carried on as in any oil mill. In such a plant, the delinting operation costs half a cent a pound, and as a by-product this lint could, and would, be sold cheap for chemical processing to rayon, film, and lacquer bases.

Thus semilint cotton would be a twofold help. What lint it did produce would not go into textiles in competition with ordinary cotton. It would, however, keep cotton acres in cotton production, and we need an annual crop of some twelve million bales, or twenty-four million acres, if the ways and means of living of a tenth of our population are not to be completely disrupted. Cotton for seed would mean cotton acres without cotton, a smart way of whipping that particular devil.

Semilint cotton also helps cottonseed. A chronic undersupply of an important commodity is almost as bad as a chronic oversupply. It encourages substitution. Cottonseed oil is our most important edible oil. It is the base of butter and lard substitutes and salad oils. If there is not enough to fill the growing demand for all these food products, these industries will turn to other vegetable oils. The increase in the peanut and soybean crops is thus another threat to the cotton grower. In the same way, cottonseed meal is a most important protein feed for cattle. Already, right in Texas, more grain sorghums and corn are being grown to make up for the lack of cottonseed meal.

This worldwide, interrelated complexity of lint and seed, of main products and by-products, of competition not only with foreign cotton and rayon and paper, but also with wood pulp and peanuts and corn, adjusts its enormously diversified elements in a normal market by the moving balance of price in the scales of cost and use. A fixed price freezes the balance; jams the whole economic mechanism; creates great, often unexpected strains in distant areas.

American cotton, held aloft at an abnormally high price controlled by all the financial power of the United States Government, is caught in a terrific economic and technological squeeze. On the one side, cheaper foreign cotton is expanding at a rate of some three million bales a year; three million bales literally taken from American cotton's old export market. On the other hand, the price of synthetic fibers comes down every year and some types of rayon are already selling actually cheaper than cotton. Paper, which also competes with cotton in bags and cartons and napkins and scores of newer uses, is already sold well below the price of cotton. At the same time, the dynamic forces of research and sales promotion are pushing both synthetics and paper much more effectively than any promotion of cotton and its uses. No wonder Garrard exclaims that it is about time something was done for cotton.

"Whether we like it or not," said Oscar Johnston to me, "the immediate future of cotton is not economic or agricultural, but political."

He should know. As president of the Delta & Pine

Land Company, he operates some fifty thousand acres in the best cotton section and he has served on both the Agricultural Adjustment Agency and the Commodity Credit Corporation. Furthermore, this short, baldish, Mississippi lawyer-planter, who talks so businesslike, is daddy of the National Cotton Council, a really new idea in self-help for cotton.

In the war on the political front, this recently formed Cotton Council occupies a position not unlike a government in exile. It is a unique organization. Set up upon a state basis, it gathers together producers, ginners, seed crushers, warehousemen, merchants, and spinners, all the elements involved in cotton. As in the Congress, each group is represented proportionally; the whole is financed by a prorata levy per bale grown, handled, or used. The Council has four postwar programs: for sales promotion of cotton products in domestic markets; for producing, processing, and marketing so as to increase quality and cut costs; for the recovery of exports; for research to develop new cotton products and improve old ones.

At the University of Texas a staff of the Council is mapping the defenses and offenses of this war. Its Adjutant-General is Dr. Simon Williams, a diffident young man, self-effacing until he gets on the subject of cotton. Then his brown eyes snap, his swarthy face lights up, and he says the most amazingly frank things. Having been trained in forestry at Syracuse University, he knows that cotton and wood are both cellulose, so he looks upon cotton's synthetic competitors without the least trace of awesome fear.

"Cy" Williams voiced these political tribulations when he said, "A political price is so utterly undependable that it undermines all plans for the future. It makes every commercial project shaky at the base. Any forward-looking research program can so easily end in a useless triumph that this uncertainty deadens imagination and saps initiative." His energy and enthusiasm belie his despondent words.

"Cotten is a cross-eyed stepchild," he went on as we sat at his little desk in the corner of a big room filled with spinning and weaving machines. "Every agricultural raw material competes with synthetic rivals in industry under a tremendous initial disadvantage. Nobody carries on research for their uses. Certainly the many small farmers cannot do it, and all the scientific work sponsored by the Government is aimed to help the producers, not the product."

"Them's harsh words," I said, laughing.

"But true," he shot back quickly. "Cotton must stand on its own feet upon a price base. If the Government withdrew the loans today, my guess is that next week the price would drop a half. At eleven cents most of cotton's competitors would fold up and none of us around here would have qualms about the future. As it is, faced with the chances of a political future, all that we can do is to try to put the fiber in a position eventually to meet price competition fairly and squarely."

The research project under Williams is three-pronged: first, to use cotton better by developing improved machinery; second, to grow cotton better and improve the quality; third, to make cotton better by modifying its properties chemically. This is self-help with a vengeance. It requires the services of twenty-five experts and costs $350,000 a year.

"That's a drop in the bucket against what a chemical company would do for one of its own new synthetic products," said Williams, "but we've made a start and we're doing some interesting things."

Down in Mexico, Council agronomists are growing two cotton crops a year, saving half the time in their plant-breeding experiments. Before the spinning and weaving operations, Council chemists are impregnating the individual cotton fibers to strengthen them, to make them water-resistant and mothproof, to give them other desirable qualities. They are spreading out cotton batting and impregnating it with plastics to make laminated sheets of astonishing strength.

Government research at the beautiful, elaborately equipped Southern Regional Laboratory, New Orleans, is also fixing its sights on industrial uses. The guns are manned by three able musketeers, Ed Gastrock, Kyle Ward, Jr., and R. J. Cheatham, captained by Director D. F. J. Lynch. The bull's-eye is industrial fabrics, the coarse, heavy stuff of tirecord, bags, tarpaulins, belting, and filter cloth, which consume nearly half the cotton we use. Much secret work has been done for the armed services, but the crimped cotton bandage, one of the developments, is now news. Lengthwise and breadthwise it has remarkable elasticity, fitting snugly over bumps and angles of the human anatomy, and, being preshrunk, it can be used under a cast; not a new application, but surely a first-class improvement.

The Southern cotton textile industry has broken out in a rash of research. Suddenly it has come to realize that the fabrics of the future will be tailor-made mixtures of all the natural and synthetic yarns. Prewar, it went along complacently spinning and weaving cotton goods. Today, it is experimenting frantically lest it be caught without the knowledge and machines to handle rayon, Celanese, nylon, as well as cotton, in all possible combinations.

Made-to-order for every use from stair carpet to summer dress, these new fabrics will give you and me textiles of undreamed beauty and durability and, above all, of greater suitability to our needs. Our clothes will be cooler and wrinkleless for summer; warmer, yet lighter in weight, for winter. Draperies will be really colorfast and fireproof. Wash goods will be made for the washing machine and the electric iron. From the consumers' point of view, the possibilities are delightful, but they open up a new phase in the competition of cotton with the synthetics.

At the new Textile School of the North Carolina State College of Engineering, Director Malcolm Campbell showed me a spool of ring-spun cotton yarn so fine that two hundred miles weigh but a pound. That kind of yarn has not been made in this country.

"American textile mills," he said to me, "have never made the finest of sheer cotton fabrics. They will after the war. There's a revolution going on in our Southern cotton textile industry and many new goods will result in upsetting our old ideas about cotton textiles. I can conceive that a few years hence women may be boasting that they are wearing sheer cotton undies, luxury fabrics made out of cotton."

All this research is forging new weapons for the cotton-synthetic war. At the same time, recognizing that it will be hard to recover cotton's lost export markets and not easy to raise domestic consumption, another college professor has proposed as wild a self-help scheme as Killough's cottonless cotton. He would toss the entire cotton plant into a chemical vat.

A dozen years ago an inquisitive chemistry student at the University of North Carolina asked a leading question, "How much cellulose is there in the cotton stalk?" This set that professor thinking. Why not harvest the whole cotton plant with a mowing machine, extract the oil with solvents, and use all, stems and stalks, bolls and lint, as a source of chemical cellulose?

Dr. Frank K. Cameron soon found that if the plants were crowded three inches apart they matured quickly in twenty-five or thirty days, and that the machine for cutting and baling hay could be quite easily adapted to the cotton plant. From there on he ran into discouragements. Not only did a whole new series of techniques have to be worked out, but many subtle influences thwarted this revolutionary idea. In Washington he was told pretty plainly to forget it. Powerful lumber industries suspected him of working against the chemical utilization of wood pulp. Cotton people were completely disinterested. But he persevered.

Troublesome colors which stained the oil had to be removed by bleaching. A Chinese girl student, W. W. Chen, learned a new method of extracting the cellulose from the whole plant most efficiently. Step by step, the new process was slowly perfected.

In 1944 when the North Carolina Planning Board took it up, things began to happen. An experimental plant has been set up at Rockingham under Nicholas Dockery, an ingenious, easygoing, former student of Cameron's. When Dockery is not managing his mother's big plantation, he is interested in cotton mills and insecticides and other chemical specialties. In one of his plants bleached pulp from "whole cotton" is being prepared in commercial quantities for a real industrial try-out by the Sylvania Industrial Corporation at Fredericksburg, Virginia.

At Rockingham it appears that an acre will produce about five thousand pounds of whole cotton a year, yielding two thousand five hundred pounds of cellulose and three hundred and fifty pounds of oil. The whole plant can be delivered at the factory at a per acre cost of $20. Processing costs are estimated to be about $23. At war prices, three hundred and fifty pounds of cottonseed oil and two thousand five hundred pounds of cellulose are worth in the neighborhood of $200, but even at normal prices, $120; so the financial prospect is pleasing. The advantages of an annual source of cellulose and of taking cotton acres out of the production of lint are as obvious and tempting as a watermelon lying on the broad highway.

At Fredericksburg, Dr. Frank Reichel, the chemist-president of Sylvania, is enthusiastic but reserved. There is no doubt that "whole cotton" makes good cellulose, suitable even for the exacting uses of photographic film and high tensile yarn for tirecord. The $64 question is, can close planting, mechanical cultivation and harvesting, efficient processing, bring the cost of this form of cellulose down to five or six cents a pound?

Once again, we are back where sooner or later every discussion of cotton lands—at price. Nobody expects cotton to vanish from the face of the earth or even from the favorable fields of our South. It is too cheap, too usable a fiber. It has distinctive properties, such as its ability to take a beating in the washtub week after week, its absorbent qualities, its great strength when wet. But more than ever before it is now plain that cotton's position must be held upon a price and quality basis.

4

Cotton's Other Crop

THIRTY ODD YEARS AGO, on a hot Texas June morning, a painfully thin, but exceedingly wiry young man walked off the campus of Baylor University. He carried in his hand a crinkly parchment scroll duly inscribed to Thomas Jefferson Harrell and in his head the quaint notion that said Harrell was worth $5,000.00 a year of any prospective employer's money. Three months later he was shoveling cottonseed into a storage bin at an annual salary that moved the decimal point precisely one digit to the left.

Twenty-five years later another scroll was presented to Thomas Jefferson Harrell. It was signed by five hundred and twenty-four of his fellow townsmen and proclaimed him Fort Worth's citizen of the year. At the banquet when this Golden Deeds Award of the Exchange Club was made, it was solemnly declared that this now thickset but still galvanic man is as modest, as useful, and as versatile as the cottonseed upon which his career has been built.

"Togie" Harrell is president of the Traders Oil Mill

Company. Besides, he is or has been president of the local Rotary Club, the Sales Managers' Club, the Chamber of Commerce, and the Y.M.C.A. Executive committeeman of Fort Worth's Show celebrating the Texas Centennial, mayor of his city, Harrell has tied a record of unusual achievement to all his many titles. But to compare his accomplishments to those of the humble cottonseed—well, Fort Worth is in Texas, and in Texas bragging is the king of local sports where the native toreadors hold all the long-distance slinging records.

"Togie" has made good. He and every other man in the business admits that. But the cottonseed—it has made an industry of some four million tons bulk worth over two hundred million dollars, out of an unmitigated nuisance. No man is that good. Cotton's other crop is one of the South's choicest assets, and except possibly the products of the pine woods, the products of the cottonseed touch more Southerners directly and beneficially than any other raw material that originates below the Mason and Dixon Line. That, too, sounds like Texas talk, but here are the corroborating facts.

It was "Togie" himself who, when he was president of the Texas Cottonseed Crushers' Association, designed a traveling display of one hundred and fifty-one different products made from cottonseed. That was back in 1929; today it would not be hard to nearly double that number. An up-to-date exhibit would show such war-born novelties as the chemical furfural for the synthetic rubber program and artificial cocoa, pepper, cloves, and cinnamon for the pantry shelf. For the synthetic condiments, Thomas Jefferson Harrell is himself indirectly responsible.

Shoveling cottonseed out of a freight car is not raking leaves on the lawn of the Public Library and today $10 for a sixty-hour week is not even coolie pay. As he wiped the sweat out of his eyes, T. J. Harrell, B.A., no doubt wondered what good a college education was to him anyway. But it did teach him to use his eyes and his brain, to see things from all sides and then think them through from an original premise to a definite conclusion.

"Togie" confesses that though he did not like his first job, he did like the crushing business from the start. From the soft feel of the fuzzy seeds and the aromatic smell of the cooking kernels to the hazards of a seasonal, speculative enterprise and the flinty competition with big and little rivals for both raw materials and sales, he just liked it all the way through. Maybe heredity pulled some strings. His grandfather, back in the 1860's, owned and operated one of the first cotton gins in Texas, a four-horsepower gin, literally four mules on a treadmill, with a daily capacity of four bales. To his grandfather, cottonseed was a bothersome nuisance. He dumped it in the river.

From the receiving platform where he shoveled incoming seed, young Harrell began to observe every operation of the oil mill. He learned that damp, dirty seed will rot and sometimes it bursts into spontaneous combustion. So when precleaning became standard practice he already knew it made good sense to dry the seed and remove twigs, leaves, bits of boll before storage in the seedhouse. He saw the seed cleaned and then pass through the delinting machines where revolving circular saws cut off the fuzz of lint and whirling brushes collected it, sending it to the baler. He observed the differences of the single pass-through and its "mill run" linters and of the double delinting with its "first cut" and its "second cut" products. He learned first-hand how the hulls of the seed are sliced in the huller, then separated from the meats or kernels, and he could appreciate the fine points of bar or disc huller and the comparative merits of all different types of shakers, beaters, and other separators.

Before he could vote, Harrell followed the hulls to storage and shipping rooms and watched the meats squeezed into flakes by steel rollers and then dumped into the big cookers. It was years later before the first steam-pressure cookers appeared with their saving in cost and improvement in yield. Pressing the cooked meats into slabs and wrapping them in filtercloth held no secrets, and he was familiar with all the tricks of stacking the slabs in steel boxes, one above the other, a dozen or fifteen high, and then squeezing them together with a powerful hydraulic ram, four thousand pounds to the inch, till the yellow oil flows out to the storage tanks, leaving the expressed cake to be broken into bits and either screened to size it or ground up to meal.

Starting with a worm's-eye view, learning every step in the conversion of cottonseed into linters, hulls, meal, and oil, as "Togie" Harrell did, is quite the fashionable way of breaking into this industry. Scores of its leaders have served the rigorous A-to-Z apprenticeship, climbing to the top from the ladder's bottom rung. Its most popular figure, Fred Pendleton of Dallas, came from bookkeeper at seventeen to the unique pinnacle of having been president of the national and two state—Texas and Oklahoma—associations of cottonseed producers. Pendleton was awarded an engraved scroll, too, on the occasion of his having "served the cottonseed industry, generously and faithfully, for half a century." Such practical, working leadership no doubt accounts for the great number of "independents," men who have never failed to discount their bills in an industry where the pitfalls to trap the unwary are many and competition is razor keen. This up-from-the-ranks tradition also accounts for some important improvements in methods and machines that have sprung from within the mills.

It was the president of an oil mill in Alabama, T. J. Kidd, who perfected the delinting machine which strips the tiny hulls of their fuzz of cotton fiber. Ben Clayton of the famous Anderson, Clayton & Company developed a method of refining crude cottonseed oil by whizzing it in a centrifugal machine, a process that recovers considerably more refined oil than was formerly possible and which today is employed by practically every refiner.

The pattern of the cottonseed-crushing industry is familiar enough in this country: a few large corporations and many small companies which enjoy peculiar advantages so that they are by no means at the mercy of their big competitors. But, from the point of view of the small crusher, this familiar competitive situation has one distinctive and disconcerting peculiarity. None of the big boys is really in the cottonseed business at all. All are primarily big consumers of cottonseed products who operate oil mills first to protect their raw material position. As the small mill owner sees it, they seem to have sketchy ideas about the costs and profits of these operations.

"They wrap up their losses in a bullhide," is how "Togie" Harrell explained it to me, "or drown them in a can of salad oil, or tuck them into a cake of soap. If the small mill man sells his oil and meal as raw materials, he is caught naked in a blizzard. He must protect himself by finding something that he can make out of his cottonseed products and sell it himself so as to earn a manufacturing profit on his own raw materials."

Twenty years ago Fred Pendleton had come to a

somewhat similar conclusion from a different premise which grew out of the conditions of the years following the First World War. At that time most of the cottonseed meal throughout the South was sold to the big fertilizer manufacturers, they used it as a premium ingredient—rich in nitrogen and organic material—in their mixed fertilizers. In the Southwest, however, three-quarters of the cottonseed meal was shipped abroad in big, flat, hard slabs. Most of this went to Holland and Denmark where it was used as feed for dairy cattle. What was left in these states was also fed to cattle, beef cattle in feedlots close to the local mill.

This was a jolly setup for the oil-mill owners and managers of the Southwest. For a few months in the late summer and autumn they worked overtime and Sundays, buying cottonseed and running their mills to capacity. Then some long distance telephone calls to sell their oil to dealers and manufacturers and their meal slabs to brokers and exporters, both at known prices fixed on the commodity exchanges in New Orleans, Chicago, and New York. Fill the tankcars and load the boxcars; sweep out the mill and the warehouse, smear the machinery with grease; lock the door and go fishing!

World War I jarred this smooth-running system. Owing to the dislocation of ocean shipping, the exports of meal slabs shrank close to the vanishing point. But they came back in the postwar boom, and most of the cottonseed men of the Southwest promptly forgot those worrisome war experiences; but not all. Pendleton was one whose memory was good.

About the same time, agronomists and chemists began saying some disconcerting things about cottonseed. The cotton plant, like the nylon factory, makes its fibers literally out of air and water. It is in the production of the oil and protein of its seed that the cotton plant draws heavily upon the fertility of the soil.

Just at this time too, the rampant boll weevil made cotton growing an extremely hazardous occupation. Agricultural experts and county agents began preaching the gospel of salvation in beef cattle and diversified crops. The slogan "Balanced Farming" became familiar.

Fred Pendleton put together these three unrelated facts: the vulnerability of the export market; the soil depletion of the cottonseed; balanced farming; and he built up a new idea which he expressed as follows:

"We can't go on burning hulls for fuel and depending on our foreign markets for meal any more than the South can go on growing cotton, alone, depending on fertilizers to make good the plantfood taken from its soil. The whole thing is a wicked combination to assure bankruptcy. When we ship meal to Europe, our own farmers lose irreparably the feeding and fertilizer values taken out of their land. We must help return those elements to the soil by cultivating local markets. The farmers and ranchers don't know us, don't like us, and know very little about our products. What can we do about it?"

Pendleton began talking these thoughts. He found several of his friends in the business had been asking themselves the same question. J. Webb Howell came up with a suggestion. He lived at Bryan, next door to Texas A. & M. College, and had a neighbor who was working for the Extension Service, a swine specialist with ideas about this balanced farming business.

"You ought to talk with young Ward," he told Pendleton. "A. L. Ward, A. & M. graduate, managed a big cotton-livestock outfit up near Paris—was a major in artillery during the war—he talks your language, talks it effectively to farmers, and gets things done. Next time he's in Dallas I'll wire you."

So one day in February, 1926, the telephone rang in Ward's room at the Baker Hotel, and half an hour later a big, expansive man walked in. Five hours later Ward almost missed the nine o'clock train back home.

That talk between Fred Pendleton and A. L. Ward reaped a bountiful harvest of deeds. The ex-major was invited to tell a small group of North Texas mill operators where cottonseed products fit into the balanced farming program. They were interested in Ward's ideas, and in Ward, who was asked to tell his story before the annual meeting of the entire Texas Cottonseed Crushers' Association at San Antonio in May. Again, Ward sold his ideas, and himself, so that before that meeting adjourned he had been offered, and accepted, a brand-new kind of job. He was to organize an "Educational Service" for the cottonseed crushers and make it effective in practical benefits to the common interests of the oil mill operators, the cotton planters, and the livestock producers.

"Texas is big, but this idea is bigger than the Lone Star State"—so Senator Christie Benet of Columbia, South Carolina, who was the banquet speaker at that convention, told the assembled Texans. He was right, and in a short time the Educational Bureau became a major function of the strong National Cottonseed Products Association. A. L. Ward is still making that program effective. It has furthered the balanced farming idea immeasurably, and today cottonseed meal is the most important protein feedstuff in the country.

"Caking," which is rancher's lingo for feeding cottonseed cake, began as a sort of tonic for sick cows. Then cattlemen of the West and Southwest accepted the heavy annual loss of weight and even deaths suffered during drought and winter months as just part of the business. Where cottonseed cake began to be used as a lifesaver for weaker cows in the breeding herd, it was administered in small lots in fenced areas, but the results were as obvious as a black Angus bull in a snow field. Smart cowmen began caking on the range. Here this rich supplemental feeding began to grow beef quicker for market, and caking became a big factor in the evolution from heavy, aged steers to the modern type of young beef animal.

When Harrell made up his mind to do some research to develop his local markets for manufacturing cottonseed products, caking did not need to be sold to the wide-awake cattlemen. But meal was still his only local product. Unless he was prepared to go into the manufacture of margarine or vegetable shortening or salad dressing or soap, he would only waste time and money on research for cottonseed oil. A scientific study of linters would be profitable work only if he planned to branch out into upholstered furniture or mattresses, rayon or lacquers. But meal and hulls were something quite different. These were sold, not to big companies which had greater research facilities than he could hope to organize, but to a lot of cattle feeders. These men were his own customers, men he might run into at lunch at the Fort Worth Club any day, men he knew as "Jim" and "Harold" and "Baldy."

"To hear those cattlemen squawk, you'd think they had corralled all the troubles in the universe," so Harrell reasoned, "and it looks like a fellow might make friends and influence a bit of business, if he found out what they wanted and gave it to them."

So out he went to the ranches. There he soon learned that the cottonseed cake he and all other millers were selling was not pleasing to the customers. The theme song of complaint was its bricklike hardness. "Your meal is pure, forty-three per cent protein stuff, all right," he was told time and again, "but say, it's fitter to feed to a granite crusher than to steers."

Back Harrell came to his mill with a live, practical research idea: find some means of controlling the consistency of cottonseed meal. He had not pondered that problem long before he reached some conclusions. The cattle gobble up the pellet form of concentrated feed most eagerly. Cattlemen like it for its easy handling, no dust and no waste. So if we can make cake soft enough, let's pellet it without any molasses binder.

Harrell did not have a monopoly on this clever idea. Out in Sweetwater, a fellow Texan, Bob Simmons, was already pelletting cottonseed meal, and there are others who claim this "first." Today pellets are made in many oil mills, and even where they are not produced, the regular cake, thanks to improved pressing techniques, is as crumbly as the crust of good corn pudding. The day of the brickbat meal cake has gone, and with it the easygoing times when the mill operators could shut up shop and go hunting and fishing six months of the year. Now the mill offices and warehouses are open the year 'round to handle local sales of feed products. Furthermore, mill owners and managers have had to learn about "white faces" and "creep feeding" and "protein values." A lot of them have taken extension courses in animal husbandry at their state agricultural colleges, and all the way from Alabama and Florida to Oklahoma and New Mexico, wherever cattle loom large on the landscape, they even ship in more meal from mills in other sections to satisfy their local customers. Caking has revamped the oil mills almost as thoroughly as it has the feed lots.

Mother Nature herself set up the frame within which the pattern of this cottonseed industry has been drawn. Produced over a wide area, cottonseed accumulated as a by-product at the cotton gins. These were scattered throughout the entire Cotton Belt, spotted in the center of a five-mile radius, just about as far as a planter can haul his seed-cotton in a wagon with a pair of mules. Within larger circles, so as to draw upon seed collected at a number of gins, are the oil mills.

During recent years the basic radius of both gin and mill has been lengthening. The mule is being hustled off the road by the motor. The five-mile limit has been raised, and then raised again, by the larger truck-drawn load—which means more miles and fewer trips. At the same time, over large areas, less and less cotton is being grown.

Wider areas and fewer bales of cotton have ganged up on the ginners, closing many of them, encouraging fewer and bigger and better operations. The 19,195 gins of 1923 have been whittled down to 12,033 in 1943, down to not more than 10,000 by the end of the war. So lint-hungry are the battling survivors that some of the more aggressive have gone out into the cotton fields in the picking season. They leave fleets of trailers to be loaded, and then drawn by their own trucks in long trains to their gins. Although as yet only a few gins do their own hauling, most farmers have their own trucks. Seven out of ten pounds of seed-cotton now move from field to gin by motorized equipment, and this has shoved out the collection radius to twenty-odd miles.

For the gin owner forced out of business by this competitive pressure, it is little comfort to know that he has been a victim of progress. Yet this is true since small-scale operations do not make for efficiency and economy, and growers are continually demanding better ginned cotton.

The average gin handles only a thousand bales a year and works from six weeks to two months. A large centralized gin can handle thirty thousand bales a season. Allied with an oil mill, as is already accomplished in more than one instance, it becomes a sizable operation that can afford modern equipment, maintain it properly, and man it with competent labor. The cost of ginning and wrapping a bale of cotton is only six dollars, not an exorbitant share of the charges against the grower. But if American cotton is to meet the open-price competition of cheap foreign growths and low-cost synthetic fibers, then every penny will count.

Within the wider orbit of the oil mills, similar forces have been pressing the same kind of progress forward. Casualties have been greater. Thirty years ago there were over eight hundred and fifty oil mills: now there are fewer than four hundred. However, each of the survivors is doing twice as much business, for the average crush has advanced from five thousand tons of seed per mill to over eleven thousand tons.

As the ginners have been hustling out into the fields to collect cotton, so the crushers have been scrambling around among the gins trying to corral as much cottonseed as possible. Competitive bidding for seed has been lively because, working day and night through a short season, an extra ton or two of seed boosts the profits and even five hundred pounds are not to be ignored.

This progress under pressure may be uncomfortable, but it is wholesome in its effects upon those who can stand the strain. With the best brains of the industry preoccupied with the problems of buying and selling, such vital matters as operating efficiency, quality of finished products, and the development of new markets in new employments for cottonseed products have been too long relegated to the background. Most of the technical progress has come from the outside, chiefly from the makers of machinery and apparatus, and almost all of the new uses of cottonseed products has originated far beyond the confines of the oil mills. This is an unwholesome state of affairs. Any industry that depends upon its suppliers and its customers to do its own research for it is like a community that depends upon neighboring towns for its water supply and its fire protection.

That is why the aggressive, alert "Togie" Harrell is so interesting and significant. He is really research-minded and stands as the epitome of many of the younger leaders among the big operators and the smaller independents, men with vision who scan the horizons eagerly.

The first research laboratory in the cottonseed industry was established away back in 1887 by the American Cotton Oil Company, and in that laboratory the roly-poly genius with the twinkling eyes and little, close-cropped Van Dyke beard, the great David Wesson, blazed the way to a truly refined cottonseed oil. Shortly before he died in 1934, this insatiable enthusiast for the cottonseed gave a little dinner party at which he fed his friends a wide variety of cottonseed foods, including croquettes and biscuits. We choked them down, for who would displease such a delightful host? Harrell is a son of that same good line. He was one of the first of the smaller independent mill operators to install a research laboratory and man it with high-grade technicians. He backed his research director, C. W. McMath, with time and money in the long research that resulted in the perfection of the first cottonseed flour to win official approval as a food product from the American Medical Association. This flour is being baked into crackers, doughnuts, cakes, enriched bread.

Just to prove again the versatility of cottonseed products, when the war cut off supplies of pepper and spices, Harrell's research department worked up a modified form of cottonseed flour that is being impregnated with the pungent aromatic principles of pepper, cinnamon, cloves, ginger, and allspice. He is actually running a black pepper factory.

"The smartest chef," he boasts, "can't distinguish any difference in flavor, and these new artificial condiments are really better than the originals for they actually have a little food value."

He is crowing on too small a stump, for the greatest achievements of this pioneering research experiment are indirect. His successes have turned others to research. Indeed, from neglect a dozen years ago, this fuzzy little seed has become quite the pampered pet. The margarine and vegetable lard companies have stepped up their scientific programs. Inspired by their president, James D. Dawson of Houston, and Ray Grisham of Abilene, the Texas Cottonseed Crushers have caught another starry-eyed vision. They have engaged an agricultural expert, C. B. Spencer, and sent him out over the long, straight Texas roads selling farmers and cattlemen the idea of more efficient, better-rounded farm management. It has nothing to do with cottonseed, directly; but these businessmen believe that an industry based on an agricultural by-product cannot but profit by a sounder, more profitable farm community. That is turning the chemurgic idea inside out. It might reap a rich harvest.

But there is a good chance that the lowly cottonseed may cease to be a by-product of the fiber plant and become a full-fledged crop in its own right, a straight competitor of the soybean and the peanut, of corn and grain sorghums. This is, of course, the idea behind Dave Killough's semilint cotton. Whatever the future may hold for cotton fiber, the demand for edible oil and protein stock-feed is greater than the supply and promises to become still greater in the years ahead. The horizons of cottonseed are wide and rosy-hued.

There is one self-confident little chemist, placed at a rare vantage point for scanning these horizons, who believes that the sky's the limit. John Leahy is director of the Cotton Research Committee of Texas, an interesting man energetically working on one of those new-style self-help-for-cotton projects. That committee is a brainchild of George Moffett of Chillicothe, who, amazingly, is the only farmer in the Texas Legislature. Senator Moffett is a real dirt farmer, a second generation cotton grower from the High Plains, one of those wholesale, thousand-acre planters who make money regardless of farm conditions. His practical idea was an agency to hook together and make sense of all the cotton research work in Texas, from plant breeding to dyeing textiles. It takes a fast-moving man with a wide-ranging mind to handle that kind of an assignment and John Leahy meets the specifications. He is shortish and baldish, as alert as a chipmunk, with the courage and tenacity of a pit bull terrier. Born on a cotton plantation in Arkansas, he was trained in chemistry at the University of Mississippi and has a life-long experience —less than forty years long—in cotton and chemistry.

Leahy believes that Texas cotton is the crux of the cotton problem—the biggest sectional crop, grown at lowest cost, short-staple in grade; formerly ninety per cent exported, now glutting our market—and he is enthusiastic for cottonseed as the solution. Naturally he likes Killough's cottonless cotton. He even talks about cotton as a grain crop, which he adds slyly is "food for thought for a lot of people." Assuredly cotton-for-seed would be a brand new crop.

"Cotton growers never have had any sense of the value of seed," Leahy protests. "They take a three-dollar-a-bale credit for it on their ginning bills and call it a day. As a result, the cotton crop has been hitched to the textile industry, and when we talk of diversifying Southern farming—why don't we begin by diversifying the cotton crop? We could learn from the soybean farmers. They have whooped up their bean till the whole country thinks it is an agricultural miracle. As a matter of fact neither soy oil nor soy protein is a whit better than cottonseed. But poor little cottonseed has always been an orphan of cotton trying to make its way in the world substituting for something else, which often isn't as good as the cottonseed product."

This aggressive chemist naturally believes in a chemical future for cottonseed, especially for the hulls which are the low-price member of the quartet of major cottonseed products, fed to cattle for roughage. Several years ago Leahy showed that from cottonseed hulls almost twice as much furfural can be extracted as from oat hulls. He was told to forget it, but when furfural became a critical material for the synthetic rubber program, the Government built an extraction plant at Memphis and commandeered one hundred and fifty thousand tons of cottonseed hulls. Results confirmed Leahy's neglected analysis. Almost half of cottonseed-hull bran consists of various pentosans, a varied group of chemicals that can be converted to fermentable sugars, solvents, plastics, even to motor fuels. Furfural (used as a solvent in refining lubricating oil, extracting resins, and in plastics) and xylose, which is wood sugar, are the best known, but even they have not yet come into their own. There are scores of others.

All this is still the chemical industry of tomorrow—maybe of day-after-tomorrow—but right now the humble cottonseed is one of the most important resources of the South. As Killough and Leahy expect, it might give Southern agriculture a new crop. It is sure to give Southern industry a lot of new factories and, as research continues, a lot of new products.

5

Oils We Eat

THE ROTARY CLUB of Kingsville, Texas, devoted its luncheon meeting on January 23, 1945, to the subject of dairying. Leading milk producers and dealers of the neighborhood were invited guests. It was a timely gathering, for the city's milk supply had been upset by war conditions. The three speakers were experts and what they said was to the point. One of them, the very impressive Dean of Agriculture at Texas A. & M. College, Charles N. Shepardson, made a statement that, considering speaker and audience, was most startling.

"As a dairyman," said Dean Shepardson to these Rotarians and dairymen, "I must confess that I would rather eat some of the improved butter substitutes than some of the low-grade butter on the market."

If this be treason—well, the crowd, dairymen and Rotarians alike, applauded that statement as roundly as the Virginia House of Burgesses applauded Patrick Henry's historic speech. And yet, but a few years ago another college man was forced by the embattled dairymen of his state to withdraw a report which merely stated the demonstrable fact that margarine sells for half the price of butter because its raw materials cost about half as much butterfat. Times do change, and possibly this only proves the differences between a statement of fact in Iowa and an expression of opinion in Texas. But the contrast struck me as Dean Shepardson waited for the applause to stop. Then he clinched his astonishing confession with an even more remarkable prophecy.

"Dairymen have depended on law to tax the competitive substitutes for butter out of the market. In the future we must depend upon taste, which is the consumer's test of a quality food product for which he is ready to pay a premium price."

Margarine is going to make one of the most exciting readjustments of the postwar period. Mars is himself a great fat-eater and when he operates on a global scale he seriously disturbs the world's fatty oil balance. Neither North America nor Europe have in the past produced sufficient fats and oils for their own needs. Both continents normally import billions of pounds of oils: coconut from the South Seas, palm from Africa, tung from China, and many others. So every modern war has precipitated a fat-and-oil crisis which is especially critical because we must have these materials in our kitchens, our explosive factories, our soap kettles, and our paint cans.

In striking a new fats balance after every war there

have always been some great shifts and switches. Since the use as food is incomparably the largest and most compelling, we face postwar readjustments in the three great oils we eat: cottonseed, soybean, and peanut, with lard and butter. The new equilibrium will involve some intersectional complications between the South and the Midwest.

Socially and politically, margarine has made friends and learned to influence people during the war. Coaxed by butter shortages and ration-point differentials into using margarine for the first time, thousands of housewives have been surprised at its goodness. This is remarkable because, owing to shortages of its own, wartime margarine, like wartime rayon and synthetic rubber, was not as good as the manufacturers knew how to produce. The flavor of margarine at its best has fooled expert butter-tasters in official tests. Dean Shepardson's opinion has been confirmed by many citizens who, a few years ago, would not have dared to believe even the testimony of their own taste. Furthermore, margarine's distressing habit of spluttering and acting queerly in the frying pan has been cured by more perfect emulsions. With this serious drawback removed, an honest, straightforward demand for the product on its own merits is rapidly building up in this country.

Because margarine is now made entirely from American farm products, it has won political friends. It was first made from oleo, refined beef fat, hence its original name of oleomargarine. During World War I, it was of necessity made of coconut oil, and of course there never was a farm bloc in Washington to support that great oil from the South Sea Islands. The gradual switch from coconut to cottonseed was completed during World War II, and lately soybean oil has been added, which enlisted farm support from the Middle West dairy states and thus split up the opposition.

That opposition has been powerful, determined, and merciless. The elder statesmen of the dairy group have followed the example of Cato, the implacable Roman Senator who ended his every speech with the ringing exhortation, "Carthage must be destroyed!" To them margarine is a dangerous upstart to be ruthlessly legislated out of existence. But some younger dairymen are beginning to wonder whether this die-hard policy is either necessary or wise.

The public is already critical of the high price of fresh milk, the drinking of which increased rapidly during the war. If the dairy industry cannot supply all of the whole milk and butter that the people will want, now that the war is over, the reaction will be sharp and may prompt more Government controls. Already a city's milk supply is spoken of as a "public utility," which gives a forward-looking farmer the cold shivers.

The restrictive laws against butter substitutes were shoved through Washington and the various state capitals without attracting general attention. But now that

American housewives are interested in margarine, the extremes to which some of these restrictive measures go appear to the consumer ridiculous or unfair and contrary to public interest.

This last fear is well founded. The basis of this type of legislation has always been, ostensibly, the protection of the consumer. The contention has always been that unless margarine is unmistakably identified by color, it would masquerade as butter, and the public would be twice cheated, in food value and in pocketbook. This position no longer has the force that caused the federal restrictive laws to be passed in 1886.

In the first place, pure-food laws are now more strict and more effectively enforced. Deliberate sophistication has become risky and is apt to be expensive. Since the old butter tub on the grocery counter has been replaced by the wrapped and branded package, it is also much more difficult.

Margarine is admittedly cheaper than butter. But there is not a scrap of evidence that it is in any way less wholesome or less nutritious. Both supply about three thousand two hundred and fifty calories per pound. Reinforced with vitamin A, which vegetable oils lack, no nutritional difference has been detected between margarine and butter in thousands of scientific dietary tests. In fact, when enriched with nine thousand units of vitamin A, which is the legal minimum, margarine can boast superiority during the winter months when the vitamin content of butter drops sometimes as low as five hundred units per pound.

As for deception, butter itself fools the public eight months of the year. Except when the cows are feeding on fresh, young grass, butter is naturally pale yellow, almost white, so it is artificially colored to simulate the rich yellow of the spring and early summer months. Yet every month in the year margarine must be sold uncolored or pay a prohibitive tax of ten cents a pound when it is colored with exactly the same dye that the dairymen may use without even being compelled to state so on their labels.

This discrimination goes further. Unbleached cottonseed oil and certain types of soybean make a margarine that has naturally a nice yellow tint. But this naturally yellow margarine cannot be sold until it is bleached as white as lard. Of course, if it looks like lard, nobody will spread it on bread. Taste, food value, lower price, do not change this stubborn fact.

This is not the whole story, but it is enough to indicate that margarine can work up a strong appeal to the American public's sense of fair play. Whoever will whip this up by emphasizing the fact that margarine is the only food that pays a discriminatory tax and coin a good slogan about keeping a healthful spread for the poor man's bread, will have the makings of a humming popular campaign.

In a political battle, votes count. To the consumer

backing that might be mustered, there can be added a million and a half cotton growers and now another million soybean and peanut farmers, the producers of margarine's raw materials. Margarine now has the most formidable array of friends that it has ever presented.

A potential market for over two billion pounds of edible fatty oils is at stake. There is every prospect of a surplus crop of soybeans and peanuts, and the Government is committed by law to support the price of both oils at ninety per cent of parity for two years after the war: the intra-agricultural battle of the century—fatty oils *vs.* butterfat—is staged to start.

If the Government follows the instructions of the Stengall amendment and does support the prices of soybeans and peanuts at anything like wartime levels during the next two years, then the Southern cotton grower may find himself sitting pretty under a price umbrella held by Uncle Samuel over the cottonseed's competitors. This will be a novel position. He will be enjoying the identical back-handed benefits under which his own foreign competitors have won away world markets from our artificially high-priced American cotton. The farce of legally controlled prices thus often cracks some pretty grisly jokes. In the case of the fatty oils it is exceedingly apt to turn into a thundering tragedy.

During the war American farmers almost doubled the output of that vital munition, the edible oils. In these days economic prophecy is a notably dangerous pastime, and yet none of the experts hesitate to foretell that our oil crops will be in excess of our domestic needs and that a trade-hungry Orient, with some neat stockpiles accumulated when shipping was unavailable, will be eager to supply all that fat-hungry Europe can possibly buy.

We must soberly determine whether the fact that our production of edible oils has almost doubled is going to make us a richer or a poorer people.

Most of the increase in our edible oils—from less than three billion to more than six billion pounds—has come from soybeans. Cottonseed has been "frozen," not by law, but by the simple fact that it is a by-product of the cotton crop which has been held down pretty rigidly by both legal and economic restrictions. Contrariwise, if the time comes when we must shrink our oil-bearing crops, this automatic production of cottonseed will throw most of the readjustment problems over the fence into the soybean fields. Some peacetime adaptation of the lend-lease program to help distressed Europe will certainly be adopted and will enable us to export edible oils in competition with cheaper oils from the South Seas, China, and Africa. But that will not appeal to the people of those lands who are hopefully eager to share the promised prosperity in the new world of peace. Once again economic controls may lead us into unexpected political and diplomatic complications. Quite aside from any such secondary effects, many direct conditions surrounding the worldwide markets in fats and oils fill this field with diplomatic mines and booby traps. It is foolish to hunt troubles which may never happen, but it is foolhardy to ignore danger signals of troubles that with forethought might be avoided.

Since most of the oils we get from plants and animals can be put to at least two of the three chief uses, food, soap, and paint, they are all more or less competitive. A manufacturer in any of these industries can quite easily switch from one raw material to another. In fact he never hesitates to do so whenever one oil becomes either too expensive or another is temptingly cheap. The fatty oils are annual crops; from season to season they vary greatly in yield. They come from all over the world and are subject to the most diverse political and economic conditions and to the wildest vagaries of weather. Therefore, the supply of even the most important oils fluctuates sharply, and as a result, prices rise and fall sensationally. Compared with the flighty gyrations among the prices of fats and oils, the stock market is as steady and sober as the old gray mare. It follows that the raw material purchases of these three great consuming industries follow the pattern of a crazy quilt.

All this does not mean that for each principal use there are not favored oils. Of necessity the paint and varnish industries must have a "drying" oil; that is, one that hardens into a smooth, tough film when exposed to the air. The more quickly it dries and the more durable the film, the more valuable the oil to the coatings manufacturers. All oils do not have drying properties. In like manner, the soap industry must have an oil that will saponify when treated with an alkali. While practically all fatty oils react in this manner, the resulting soap will differ greatly in all of its useful properties, depending upon the oil or fat employed. As food, the American public demands a bland oil, virtually tasteless, certainly one without any strong, distinctive flavor or odor.

Because of their particular characteristics, certain oils have become dominant in each field. These industries thus evade competitive bidding against each other for their supplies. Linseed, for example, is the great paint and varnish oil. It is not the best drying oil—tung, oiticica, and dehydrated castor oil, a newcomer born in the chemical laboratories, are all better—but linseed is a good drying oil, available in the enormous quantities that the coating industries require. Indeed it is the only drying oil that we produce in great quantities in the United States. Our domestic crop is some twelve million bushels of flaxseed, against imports, chiefly from the Argentine and Russia, of roughly sixteen million bushels.

Soap makers are not so fussy in their requirements. Into their kettles they dump tallow; low-grade animal fats, rendered as by-products of the packing industry; olive oil "foots," the residue of the edible olive oil industry; sundry fish oils. Their great vegetable oils are palm and off-grade coconut.

The food industries, on the other hand, must be exceedingly discriminating. Their favorites are cottonseed and peanut, imported coconut and olive, the great bland oils of the world. In the past dozen years, a newcomer from China, soybean oil, has stepped into both paint and food. Thereby hangs a tale whose plot in the next few years is going to center around margarine.

Soybean is an extraordinary oil with a number of uses and a lot of influential friends. But with all its advantages it could not long survive an artificially high price. Normally it is the cheapest of the four great domestic oils—cottonseed, soybean, linseed, and peanut—but not so cheap as imported coconut. This is a practical expression of comparative values and these cannot be changed by an Act of Congress. They might be changed by research that improved soy's properties or found new uses for it.

Both as food and in paints soybean has definite handicaps. Compared with peanut and cottonseed it is not strictly speaking a bland oil. It can be rendered tasteless by refining, which costs money. In time, however, its distinctive flavor returns, which is a serious drawback to its use in shortening, margarine, salad dressings, and other foods. Chemists, notably at the

Northern Regional Laboratory of the Department of Agriculture, are eagerly hunting for the secret to better flavor stability.

Soybean is not without its advantages in this field. Though helpful, none of these are inherent in the oil itself. Soybean is the cheapest of the edible oils but this substantial asset is canceled, more or less, by the greater costs of refining for human consumption and by the multiplicity of grades based upon quality. Soybean oil's best argument is its acceptance by the American housewife. She believes quite implicitly that all that saves the poor rice-fed Chinese from beri-beri, scurvy, leprosy, and starvation are the stores of proteins and vitamins lurking within the versatile soybean. If her family do not like its flavor, she is convinced that they ought to. This tribute to the publicity that has been put behind the soybean is not fanciful. The trained interviewers of a great advertising agency, polling housewives for a client that makes a famous brand of shortening, collected just such answers by the hundreds.

On the other hand, this same poll revealed a prejudice against cottonseed oil. That is not so easy to explain. Judged by nutritional values there is not a crooked sixpence difference in the two oils. The most notable difference revealed by chemical analysis is that cottonseed has more linoleic acid, which is the distinctive constituent of all the truly bland oils, while soybean oil has more linolenic acid, which gives drying properties. In the matters that really count in lard and butter substitutes the odds are all on cottonseed oil. The manufacturers who process the oils we eat agree that cottonseed makes the finest margarine, the best shortening, the most desirable salad oil. It has no strong flavor and it never develops an objectionable taste even in mixtures like mayonnaise or when baked into pies or doughnuts. This makes for a long "shelf life" which the grocer appreciates. In the frying pan, too, cottonseed behaves better than soybean. When heated an oil evidences the first signs of decomposition by smoking, and this smoke does not improve the taste of fried food. The "smoke point" of cottonseed oil is eighteen degrees higher than soybean oil. Though badly founded, the housewife's prejudice against cottonseed products is apparently real. My friend George Huhn, who had a major part in that poll, traces it back to the time when cottonseed oil, then pretty crude stuff, was nothing more nor less than a substitute for olive oil. It was a poor substitute, but it was cheap. So the inferiority complex was fixed.

Originally soybean was introduced as a paint oil. Though famous for its non-yellowing qualities in white coatings, it has never supplied more than five per cent of the oil used in paints, linoleums, oilcloth, and printing inks, the great consumers of drying oils. Today almost all—more than ninety-five per cent—of our soybean oil goes into edible products. Here most of it—

some five hundred million pounds yearly—is made into shortening where its unstable flavor and higher costs of hardening by the hydrogenation process are offset by ease and cheapness of bleaching, the fine white color of the finished product, and acceptance by the public. All these comparative merits and demerits are summed up in price, soybean selling regularly in a free market about three cents cheaper than linseed and a cent below cottonseed.

Except in wartime and to meet some critical war necessity, only a demented price fixer would dare to violate the established price ratios between such closely competitive commodities as the edible oils. The ceiling prices of cottonseed and soybean have been respectively 2.2 and 2.5 times the prewar average, but the War Production Board price on peanuts is a three-times raise above old levels. Peanuts are admittedly in a class by themselves. The oil is the highest priced of the group, and unlike soybean and cottonseed, it is not a principal product, but more or less a sideline.

For twenty years the peanut crop has been growing steadily. Since 1940 it has jumped far beyond a comparatively restricted area in eastern Virginia and the Carolinas to Georgia and Alabama and way out to Texas. The acreage in Georgia, for example, has trebled, planted largely to the small, tight-skinned Spanish type favored by the oil crushers and which most of the new farmers in Texas have cultivated. Big Virginia runners, the "circus peanut" that butter makers prefer and the cocktail and candy trades virtually demand, hold their own in the Southeast.

The peanut boom is war-born, due chiefly to the about-face from Agricultural Adjustment Agency quotas, strictly enforced, to official encouragement, flamboyantly publicized. While the peanut crop will be cut back, nevertheless it may well persist in Georgia and Alabama where diversification from cotton is no longer a virtue but a necessity and where the hard-pressed cottonseed crushers will encourage it.

Since cottonseed is the tail of the cotton dog, it is unlikely that its production will increase. By the same token, however, it will not be cut any faster than the reduction in cotton acreage. A by-product surplus is always a nasty rival for any major product, and this rigid, obligatory competition faces the soybean growers unless agronomist Dave Killough comes through with a perfected strain of his cottonless cotton plants. If Texas starts deliberately growing cotton for seed, then the South gets a new crop and the soybean farmer a new kind of competition.

One other quantity in the edible oils equation might upset all calculations. If margarine is freed of its legal shackles, we will face an oil shortage, not a surplus. From the point of view of the South this is greatly to be desired. It is greatly dreaded in the dairy sections. Striking the balance of these hopes and fears over the fulcrum of national interest, the scales slant sharply toward margarine. The basic fact that cannot be argued away or laughed off is that edible oils can be produced at half the cost of butterfat.

Before the milkmen enter the margarine battle they should consider whether they have not been more worried over the prospect of a loss of part of their butter market than the realities warrant. Butterfat is the least profitable item on the dairy industry's sales ledger, and the country is wanting more and more milk (whole, condensed, and powdered) and more cheese and ice cream. Americans might well consume—so the dieticians urge—at least a fifth more of these good foods. That fifth would more than offset butter losses to margarine. The dairyman even has one cogent reason to encourage the edible oils. Both cottonseed and soybean yield a joint product that he needs, high-grade protein feed. This, the most important, costly material that he must buy month after month, would then become cheaper.

For this country to be dependent upon imported oils is as improvident and dangerous as to rely upon foreign sources for any other essential that we can produce at home. Whether the vegetable oils have been our greatest agricultural scandal or our finest farm opportunity, the position of independence achieved during the war ought not to be permitted to slip away from us. Our sorry experience with cotton warns that our ability to hold this oil autonomy will depend upon our national policy—whether we attempt to hold the prices of vegetable oils above the natural levels of their economic value in open competition in world trade. Have we learned anything from our expensive cotton lessons? The future of American edible oils in the South and Midwest will be shaped by the reply to that question.

6

Tung—Old Oil for New Needs

FORTY YEARS AGO the beautiful upper valley of the Yangtze River was not exactly a tourist's paradise. In this mountainous back country Chinese bandit bands practiced their ancient profession right into the crooked streets of Hankow. Here they fought pitched battles for control of this rich commercial center. A white man was as great a rarity as a Mandarin in full peacock regalia in a Broadway night club and just as fair game. Alive, he was worth some sort of ransom; dead, he would tell no tales.

Into this rugged region the United States Department of Agriculture sent one of its most resourceful plant explorers, David Fairchild, a son-in-law of the inventor of the telephone, Alexander Graham Bell. Among the fruits of this daring plant hunt Dr. Fairchild brought out two hundred pounds of the nuts of the tung tree. He believed they would give us a new crop as useful as Spanish peanuts. He did not know this importation would bring with it as many problems as the immigrant English sparrow.

TUNG—OLD OIL FOR NEW NEEDS

After thirty years of costly trying we have naturalized this remarkable tree. Its cultivation will be a postwar project in which a lot of Americans will make, or lose, a lot of dollars.

The coral-veined blossoms of a Chinese tung tree in full bloom are startlingly beautiful. But it will not become a rival of dogwood and flowering crab in our suburban front yards, and you will not find it in your favorite nursery catalogue.

The tung tree bears big, attractive nuts, the size of a hen's egg. But you cannot eat them, for they contain more than a trace of a poisonous alkaloid akin to morphine. A few inquisitive G.I.'s stationed at Camp Blanding in Florida, in the region where the tung tree has been naturalized, have landed in the hospital after sampling the tempting nuts they find growing in what are obviously orchard rows. The taste discourages such gastronomic experiments. This is fortunate since a couple of the nuts would probably prove fatal.

These nuts are rich in an exceptional oil. It is a cataclysmic cathartic, so it will not make a nice salad dressing or a new butter substitute. In industry, however, it is as useful as coal tar or petroleum.

Since before the oldest records of Chinese history, the peasants of the Yangtze Valley have collected tung nuts, removed the seeds, pressed out the oil, and sold it to their fellow countrymen for many uses. For centuries this trade has centered in the city of Hankow. It was a flourishing business back in 1275 when the Venetian adventurer, Marco Polo, was hobnobbing with the Celestial Emperor. It was already an old Chinese custom, he tells us, to coat the seams of their junks with a mixture of lime and chopped hemp kneaded together with tung oil. The origins of the tung trade simply fade out into misty obscurity somewhere about the year 3000 B.C.

Tung oil is the neat trick by which the Chinese make their paper umbrellas and the paper walls of their houses tough and waterproof. It is the old, well-kept secret of those marvelous antique Chinese lacquers. Nobody has ever duplicated their depth, rich colors, or their ability to withstand heat and moisture. We keep bowls finished with these Chinese lacquers in our art museums; in the homes of wealthy Chinese they serve boiled rice in them. But we have put tung oil to work in ways the Chinese never imagined. We use it, for example, in the finest grades of linoleum and oilcloth; in the brake bands of our cars; in quick-drying printing inks; in safer coverings that insulate electric wires.

The identical properties that made it the priceless ingredient of the Chinese lacquers have won it the reputation of being the queen of all drying oils. That means that tung oil dries very quickly into the smoothest, toughest, most water-resistant film, the ideal oil for paints and varnish. So if you must have an honest-Injun, waterproof spar varnish, the kind that finishes a table top so that a wet highball glass will not leave a white ring, the kind that the United States Navy demands for masts and decks and other wood exposed to salt water, then you must get a varnish made with tung oil. For such purposes there is nothing else as good.

Thus tung oil, in its way, is as vital a war munition as rubber or quinine. It is also big business in world trade—some twenty million dollars a year or three hundred million pounds, one hundred and twenty of which we need—all previously supplied by China. When the Japs moved in, tung became a chronic scarcity. But we are on the way to declaring our new tung oil independence.

The tung nuts David Fairchild brought back from the Yangtze Valley were sprouted in California and during 1906 the seedlings were widely distributed throughout the South. Nobody burned with enthusiasm for the new tree, chiefly because American varnish makers, accustomed to linseed oil, had not yet discovered the virtues of tung oil.

Five Fairchild seedlings were sent to the superintendent of the cemetery at Tallahassee, Florida. He was not interested. The sickly looking little plants arrived on a particularly busy day. Besides, what good was a Chinese tung tree in a cemetery famous for its myrtle and magnolias? So he threw them out on the greenhouse trash heap.

A couple of nights later his horticultural crony, Captain William H. Raines, dropped in and the superintendent told about the five tung tree seedlings. Raines jumped to his feet, all excitement, for in China he had seen the stately tung tree showered with the pink glory of its blossoms. He must rescue those dying seedlings. His hobby was his garden of curious plants to remind him of the strange lands he had visited when as shipmaster he sailed the tropic seas. He got a lantern and hurried to the trash heap at once.

Before going to bed, the old sailor tenderly disentangled the dry rootlets and placed each drooping plant in a tumbler of cool water. Before breakfast he visited his patients. Two were unmistakably dead. The others, he carefully planted. One came through to become literally the daddy of the American tung industry.

This sole survivor, now famous as the "Raines tree," was no doubt a sturdy plant. As luck would have it, it was a notably prolific specimen. Like many of the great ones of earth, it has come to a tragic end. After Raines' death, it was split by a storm, but a young horticulturist from Washington bolted it together and it lived. Then the Raines house burned and the tree was badly singed. It was nursed back to bearing by another tung-tree lover, B. F. Williamson, whose pioneer nursery supplied most of the selected stock that planted the early groves. Then the bulldozer of the state highway department came along to straighten the road, routed out the historic landmark, and dragged it away to be cremated.

To come back to old Cap'n Bill and his famous tung tree in the days of its lusty youth—in 1912 he gave a hundred of his seedlings to another gardening friend, Dr. Tennant Ronalds of Live Oaks Plantation, Tallahassee. This was the genesis of the first tung grove in Florida. Next year, in an effort to reawaken official interest in tung oil, Raines sent a bushel of unhulled nuts to Washington, the first bushel collected in this country. In 1914 he shipped a bushel of hulled nuts to L. P. Nemzek of the Paint Manufacturers' Association at Gillsboro, New Jersey. From these seeds about two gallons of tung oil were pressed, the first produced in America.

During these seven years, by the cruel test of natural elimination, the only Fairchild seedlings that had survived were growing east of Houston, in the extreme southern portions of Louisiana, Mississippi, Alabama, and Georgia, and across the narrow upper arm of northern Florida. This broadside experiment proved two important points. The tung tree does not tolerate alkaline soil. In the United States, as in China, it flourishes best along the 30th degree of North Latitude. Nobody then put two and two together, and both these simple lessons had to be relearned.

At the same time, natural varnish gums were becoming scarce and high-priced. Chemists began studying the old rule-o'-thumb operations of the varnish kettle. Synthetic resins began to appear. Imports of tung oil were climbing, and as its virtues came to be appreciated, the vices of the Chinese product—uncertain quality and fluctuating price—became painfully apparent.

Some people began to think that Dr. Fairchild had something after all when he introduced the tung tree, and Dr. Ronalds soon had plenty of rivals. Spurred by business motives, two hundred paint makers chipped in $1,000 each for a demonstration planting of two hundred and twenty-five acres. L. P. Moore of the Benj. Moore paints became a real enthusiast. He backed a grove and the first tung-oil pressing mill, a costly but enlightening experience. At this point three real trail blazers set out their first baby tung trees.

Harry W. Bennett met the tung tree in its native China. He and his brothers sold their Wheatsworth cracker to the National Biscuit Company for three million. After putting through such a deal, one needs a vacation: Harry took a journey 'round the world. He was not deliberately hunting a hobby, but he found one. He returned a confirmed tung enthusiast at sixty-five. Today at seventy-nine, he is whipping up interest among fellow growers to organize a cooperative through which they can pool their fertilizer orders; establish trade standards and stabilize prices; carry forward scientific research in tree breeding and cultivation and better methods of oil recovery. When he was active in business Bennett baked a dog biscuit shaped like a bone. It did not fool Fido. It did beguile dog owners into buying this novel-shaped biscuit, sized shrewdly to suit all breeds from Pomeranian to Great Dane. As a tung grower, when Mother Nature balked, Bennett studied her whims and actually persuaded her to co-operate in encouraging a Chinese tree to settle down happily in Florida.

Bennett planted his thousandth acre in 1930: the first thousand of any single owner. He coaxed the Florida Experiment Station authorities into making the first scientific studies of tung culture on thirty plots of two acres each which he donated to the cause. Early and late he has proselyted for tung and three of the biggest, most successful growers are his converts. Scores of little landowners owe their tung groves to his initial preaching and his continuing, friendly, expert advice. Other pioneers have fought to establish the tung industry in Louisiana and Mississippi and Georgia, but none has a longer and more distinguished record than this adopted son of Florida.

Westward near Picayune, Louisiana, two big lumbermen, Lamont Rowlands and L. O. Crosby, had caught the tung fever, sensing an opportunity to reclaim their cutover timberlands. Rowlands, a tall, good-looking, quiet-spoken Michigander, cherishes a deep sense of past obligation to the denuded acres that fed his sawmills. With the same ideal of new values for Southern land, Crosby has worked out different conclusions. To a ripe old age, this tower of energy, a big man with the chest of a prizefighter and a fighter's spirit, has battled for rosin and turpentine, Satsuma oranges, and tung oil, crops to diversify local agriculture. A stout believer in the small farmer on small holdings, he planted tung trees on thirty-five hundred acres of cutover land and sold quarter sections all in good faith. When these experiments failed he shouldered the responsibility, going back and pulling the pine stumps which he fed to his wood-rosin factory to help make good his personal loss.

About the same time serious work on increasing the nut harvest began in Florida where, at the University, Dr. Charles E. Abbott made a painstaking study of the fruit bud development. B. F. Williamson of Gainesville, to whom the tung industry in this country owes much, began his first careful selection of seeds from the most fruitful trees. He started one of the pioneer nurseries which sold young stock not only in this country but to South America, Africa, and other 'round-the-world points. During its infancy, Dr. Henry A. Gardiner of Washington, possibly our first authority on paint oils, did a great deal to instruct the coatings industry on the usefulness of tung, and Charles C. Concannon, then head of the Chemical Division of the Department of Commerce, was indeed a friend in need.

The pioneers learned the fussy habits of the tung

tree the hard way. Because the Chinese harvested the nuts from semiwild trees and what planting they did was haphazard, it was concluded that tung might be grown on marginal land with little or no cultivation. The idea was to reclaim big acreages by setting tung trees out among the pine stumps on uncleared, uncultivated land. Rowlands sent an observer to China and he had confirmed this misconception. Accordingly, Rowlands started this wholesale reclaiming. In 1934 when I first visited his plantation—Rowlands then had some eight thousand acres—he was planting cleared land in contours, that is, in rows following the natural curves of the land. However, he still believed that fertilization was positively harmful.

Through many such trials and errors the pioneers worked out successful methods of cultivation. If the tung tree is to flourish, it must be planted in neutral, or better, slightly acid soil. If it is to bear commercial yields, it needs to be cultivated and fertilized. If its roots reach the subsoil water, it sickens and dies of "wet feet," so it must be planted twenty feet above the "water table." It likes a home on the hillside, which assures good drainage and air circulation, which help to prevent frost damage. That hazard of frost always haunts the tung grower. The optimists hope to get four full crops to one frost-nipped failure. The realists plan to give Mark Twain the lie and "do something about the weather."

The ruddy-faced octogenarian, Harry Bennett, again steps forth spryly as leader of the frost fighters. During the 1939 bloom, when his radio warned him of a killing frost, he flew a plane low over his groves, stirring up the currents so that the cold air could not settle. He saved his crop amid heavy losses among his neighbors. He could not stave off the 1943 frost this way because the Army, reasonably enough, had prohibited night flying by civilian planes; but Bennett has other ideas.

"Why can't we lick the weather?" he asked me defiantly. We were breakfasting in the cheery, bright-colored dining room of his modernistic bungalow at Tung Acres. He drained his third big cup of black coffee and went on, "The plant experts are breeding trees that bear more nuts and are more resistant to frost. This will help. Why not steal an idea from the Chemical Warfare Service and use some of their smoke screens to dispel frost? What's the matter with spraying the blossoms with a wax solution or some other chemical frost preventive?" Having asked these questions, Bennett is backing experiments to find the answers.

"What a healthy, happy, prolific tung tree needs is no longer a matter of by guess and by gosh," summed up this energetic veteran. "All that expensive, discouraging trying is done; proved in the grove. More nuts per tree is also a demonstrated fact. Twenty years ago, one thousand pounds of nuts per acre was good; today, the target is a ton."

Five years ago, the Department of Agriculture launched a double-barreled campaign for tung nuts: the first, purely agricultural; second, industrial, experiments on quality, properties, uses, etc. The project is under the direction of Dr. George F. Potter, a capital "nutologist," who makes his headquarters at Bogalusa, Louisiana. Three additional laboratories have been established: at Fairhope, Alabama; Cairo, Georgia; and Gainesville, Florida. Cooperative work is carried on with the Mississippi and Florida Experiment Stations, as well as with growers located in all areas of the tung belt. The scientific work at Gainesville is carried on in a modest little laboratory under the hill on the University of Florida's beautiful campus. Here a New Hampshire born, French-Canadian bred horticulturist, Dr. Felix Lagasse, and his staff have taken new steps toward greater crops, studies on breeding, selection, budding, fertilizing, cover-cropping systems, and cultural management.

In 1938, a microscopic hunt was carried on throughout the tung belt to find the most promising tung trees. Some of these trees appear to be significantly better than average tung seedlings. If this continues it will be of much economic value to the industry. Seeds from these trees were planted in 1939 and the seedlings went into the orchard in 1940. The first fruit crop was garnered in 1942, and the trees have been thoroughly tested for their capabilities to date.

A similar hunt for extra-good trees has been made by the Florida Experiment Station and they have nine, christened F-1 to F-9, that are most promising. Two of them—"sports" as the breeder calls a freak of nature which reproduces itself—are outstanding, F-2 and F-9. Both came from Thomasville, Georgia, and are said to produce almost ten times the crop of the average tree. These exceptional specimens are used for bud grafting on the roots of ordinary stock.

At a neighboring plantation, bigger yields both of nuts per tree and of oil per nut is getting a lot of attention from Robert Essa and James Wershow, locally known as "the chemical boys from Yale." This pair of robust two-hundred-pounders manage the five thousand acres owned by M. B. Jasspon. A native Georgian, Jasspon spent his early life in family enterprises in Southern commodities, cottonseed oil and meal, raw sugar, and lumber, and is now vice president in charge of new projects for the Commercial Solvents Corporation. Hence, no doubt, the scientific approach.

"We take nothing for granted," says Jasspon, "and we are not afraid to walk off the beaten path. For example, we are all for the low-branching trees, pruned like a peach tree. From selected test trees on three acres, we have results that promise three tons to the acre, not one ton."

Science has also touched oil-processing methods. At the oil mill, the hulls are removed mechanically; the nuts are broken in a ball mill; the cracked meats pressed cold. Of the four hundred pounds of oil in a ton of nuts, some seventy-five are lost in hulling, milling, and incomplete extraction by pressure. Extraction of the oil by means of solvents has been tried, and though the eleven millowners naturally pooh-pooh such a newfangled notion, it may well win out in the end.

Jasspon's young chemists are experimenting with acetone and methanol (wood alcohol), securing ninety per cent yields of oil against eighty per cent by pressing. This seems worth while. As a sort of sideline experiment these same chemists have isolated the poisonous substance from the nut meats and more of it from the leaves. One blessing of the tung grower is that his trees are almost immune to attack by insects. Hopes are raised from this immunity that this alkaloid may prove a cheap, potent agricultural insecticide.

There are now some four thousand American tung groves, ranging in size from a few dozen to ten thousand acres. Three-quarters of our tung oil comes from planters with more than five hundred acres. So tung is big business, and many of the groves are owned by really big businessmen.

Besides the lumbermen, Rowlands and Crosby, the retired Bennett and the very active executive, Jasspon, other important tung people are Charlie Goodyear of the Southern Lumber Company; Everett Paul Larsh, Dayton manufacturer of electric motors; and Roy Moffatt, Chicago machinery maker. The General Tung Corporation, started by Randall Chase, has as its biggest shareholder Carter Carnegie, nephew of the steelmaster. In Alabama the largest grower is Earl Wallis, the Chicago advertising man. United Fruit, Newport Industries, Ford Motors, and several paint companies represent corporations seriously interested in this new agricultural field.

Will tung oil remain large-scale industrialized agriculture or will the small-acre, subsistence farmer become its mainstay?

Pick your own answer from the opinions of the authorities. J. A. LaFortune of Amarillo, Texas, past football player and present trustee of Notre Dame, whose thirty-eight thousand trees at Lucedale, Mississippi, are managed by his brother, has positive ideas.

"The tung tree, like the steam yacht and the chorus girl, is a rich man's plaything. At least it is today," so LaFortune believes. "Under present conditions land values are inflated. Labor is so scarce and so handsomely paid that the average Negro in our section, making $2 to $2.75 a day, doesn't want to work more than three days a week. Bearing groves bring fancy wartime prices. New-planted groves won't come into production for six or seven years. With the ever present chance of frost, only a well-heeled man dare take such risks."

On the other side—there is no compromise among

opposing opinions—Edward C. Gray of the Gulfport Seed Company, an old turpentine and lumberman, who for ten years has been in closest contact with many tung growers, big and little, believes the future rests with the small producer.

"Tung trees need cultivation and fertilizing, but," Gray reasons, "a farm family can handle one hundred tung acres efficiently and with little labor expense. The most economic farm unit throughout the tung-growing section is five hundred acres. The small farmer who apes the big grower and tries to live by tung alone takes unreasonable chances. Half of his land, anyway, is unsuitable for groves, and if he uses this half intelligently, he can sit snugly on a three-legged stool—tung oil, beef cattle, and legumes for cover crop and feed grown between the terraced tree rows. To help initial expenses he always has lumber, for the land must be cleared. On such a farm, he sits pretty."

Another question looms large before our youthful industry: the future price of oil. As always nowadays, future profit estimates are made hazardous by three glaring uncertainties: the value of the dollar, the wage scale, and the cut that taxes will take from income. But here there is greater unanimity among tung men.

The prewar average price was just under fifteen cents a pound. Despite higher costs the postwar price will have to be close to that figure. Though some growers are inclined to laugh off Chinese competition, the

Chinese have become export-conscious. They do not intend to lose the tung trade to us as they did tea to Ceylon and India. Under the Ministry of Finance a tung oil research institute has been set up with a five-year plan for experiments in breeding, cultivating, and processing. Maybe American tung oil will not have a monopoly on its advantages of lighter color and more uniform quality. At all events Chinese competition must be reckoned with.

Right at home, tung wisemen know that if the price stays too high, chemists will be tempted to modify cheaper oils to give them desirable tung oil characteristics. Such work as has been done by A. G. H. Reimold, president of the Woburn Chemical Company, and his chemist, K. A. Pelikan, in treating castor oil to make it tunglike, will continue. Farsighted tung growers keep an eye on experimental plantings of castor beans in Texas and Florida.

Newer and more threatening is the prospect of competition from synthetic chemicals. One of those startling surprises that chemists spring upon us is that a white powder, which is really an alcohol, can be combined with rosin and oils to make an excellent drying oil for natural and synthetic resins. This contradictory chemical is pentaerythritol. It again surprises us, being the base of the superexplosive PETN, so powerful and touchy that before it is loaded into bazooka rockets and similar missiles it must be diluted with TNT. Before the war, when it was a comparatively rare chemical, the drying oil prepared from it sold for fifteen cents a pound. During the war pentaerythritol was made in vast quantities and its cost greatly lowered, so that the prospective competitor of tung oil prepared from it can be expected to be cheaper in price and available in any desired quantity; ready, in other words, to serve as a very effective brake upon runaway prices for any of the natural oils used in modern coatings.

The wartime price of thirty-nine cents for tung oil is away out of line with peacetime values. It was based on the $100-a-ton price for nuts fixed by the Commodity Credit Corporation. Jasspon says, "Growers can make satisfactory profits out of nuts at $50 a ton, and $100 is an atrocity that promises outrageous inflation of land values. Good Florida farm land that I bought three years ago for $25 an acre cannot be touched today for $50. The discovery of petroleum in the neighborhood has added to the inflationary values, but that high subsidy price on tung nuts plays directly into the hands of the land speculators."

Once upon a time an orange grove was touted as the easy way to security. The real estate columns among the want ads in Northern Sunday newspapers were bejeweled with literary gems describing the salubrious climate, the easy labors, the rich profits that awaited the lucky owner of the famous El Succero Orange Groves of twenty acres of two-year-old trees offered for sale at the ridiculous price of $224.98 an acre. After the orange grove racket played out came the pecan groves, and after the pecans, the promoters took up tung groves. That was fifteen years ago and the tung boom was killed by the collapse of tung-oil prices during the depression. But now it seems distressingly likely that a favorite postwar project of foxy real estate promoters will be the tung grove.

The saddest part of this sorry business is that the bait of a tung grove in the Sunny South will be most alluring to two groups who can ill afford such a risk: the elderly war worker and the returning veteran. To both an escape to farm life will be a powerful magnet. It is a favorite delusion of the city man that an orchard is a sort of agricultural savings bank, managed by Mother Nature, where deposits are guaranteed by God. He pictures himself as an orchardman, cashing his interest checks in the form of a profitable, annual crop. The combined charm of what appears to be the safest and the easiest of all agricultural activities in the pleasantest of climates is just about irresistible.

Orange and pecans are substantial Southern crops; tung oil is now an established product. Each is a highly specialized farm business. The men who succeed are those who play the game according to the rules laid down in the grove by the trees and in the market by their fruits.

After thirty years of trying, we have over four million

bearing tung trees. Planters expect year by year to better the record 1942 crop of five million pounds of oil. In a word, growing tung nuts and processing the oil is now established here. The business has been stimulated by war conditions. A supersecret war use may well sop up every last gallon of American tung oil, if electronics live up to half their promises. So the prospects for the postwar period are admittedly brilliant. It is an opportunity for any man with some risk capital and a liking for farming to grow an industrial crop that competes with no other in this country. Best of all it is an opportunity to diversify agriculture in the South.

And as that oldtime, practical tung man, Albertus Miller, puts it: "We welcome more tung growers, but what the American tung industry needs is not flighty speculation, but solid promotion."

7

Naval Stores Become Chemicals

GENERATIONS of Northern visitors in the South have thrilled at their first sight of a primitive little still, roofed over with some rough pine planks on four posts and surrounded by a litter of wooden barrels. "Aha," they think, "genuine 'white mule' in the making; a real moonshiner's set-up!"

They are romantically wrong. That is a chemical plant, a representative of the oldest American industry, an industry distinctively of the South. Theirs is a forgivable mistake, for that little iron retort with its twisted copper neck is also a still. However, into it is fed not a fermented mash of ground cornmeal, but a sticky, pungent, grayish gum collected from pine trees. The distilled liquid that drips from the end of its wormlike condenser is not corn whisky, but turpentine.

This elementary chemical business was started by Captain John Smith of Jamestown, Virginia. The Commodity Credit Corporation all but ended the days of the little backwoods "turps still."

By peremptory decree to assure supplies for her

Royal Navy, Queen Elizabeth, with a flourish of her goosequill pen, encouraged the production of pitch, rosin, and turpentine to become the first export business of the American colonists and gave these products of the Southern pine forest their nickname of "naval stores."

That brash soldier of fortune whose bewhiskered head was saved by Pocahontas, was a great braggart, but he never set up to be either scholar or economist. Nevertheless, in 1609, he foresaw clearly the troubles hiding in this business of making naval stores. From birth this enterprise needed the stimulus of royal favor; during the past thirty years it has repeatedly required the beneficent attentions of the United States Government to keep it healthy.

Those shrewd English merchants, the Honorable Gentlemen of the Worshipful London Company, who financially backed the colony at Jamestown, optimistically hoped to find gold; but realistically, they planned to develop a substantial trade in raw materials. Accordingly, they sent over eight Poles and Germans, skilled in making wood tar and pitch, glass and potashes, and continually exhorted their manager on the spot, Captain John Smith, to ship them profitable cargoes. When they pressed the point, the doughty Captain wrote the treasurer in London, setting forth his views on these industrial enterprises in no uncertain language:

"But if you rightly consider what an infinite toil it

is in Russia and Swethland, where the woods are proper for naught else, and though there be the help both of man and beast in the ancient commonwealth which many a hundred years ago has used it; yet thousands of those poor people can scarce get the necessities to live by from hand to mouth."

Undoubtedly the old swashbuckler had a very feeble business instinct. He would much prefer to go exploring than laboriously to sweat tar out of pinewood. However, his wordy excuses were prophetic. Throughout the colonial era, and afterward, the production of pitch and tar, rosin and turpentine grew until American naval stores dominated world markets. As a gray-bearded Georgian, veteran of many seasons of "turps farming," phrases it: "It alluz was a bizness fuller of troubles than dollars."

When a live pine tree is wounded it secretes a healing, protective gum. By chipping the bark in a succession of V-shaped cuts, one above the other, the flow of this rich, aromatic oleoresin is stimulated and continued so that it can be gathered in cups fastened beneath to the trunk. From the Carolinas to Louisiana some fifteen hundred Americans engage in this enterprise between March and August. Some work their own trees as a sideline of farming; others are professional operators who rent "gum rights" on the basis of a "crop" of ten thousand "cut faces." The work is done by some twenty thousand Negroes who make the rounds chipping the trees and gathering the gum from the cups. By the time the crude gum is collected it has been diluted with rain water and contaminated with bark, wood chips, and pine needles. Its quality varies almost infinitely with the skill of the chipper and the scruples of the operator. At best it is not an ideal raw material for a delicate distilling operation.

Hundreds of those little "fire-stills" process this gum into turpentine and rosin. These primitive chemical plants are operated without even a thermometer, literally by ear. The colored still-man judges the temperature by the sound of the bubbling in the worm and shuts down when he reckons the batch is done. What is left in the pot is strained off into barrels and hardens into the rocklike rosin. The turpentine and rosin from these many little operations are about as uniform as a flock of barnyard chickens.

All the tangible assets of the whole naval stores industry are not such as tempt a banker to lend money. Yet the small farmer and little still owner cannot carry themselves until their rosin and turpentine are finally converted into cash by sale to the consumer. Here the factor stepped in. He advanced money on the future crop at eight per cent interest, on condition that the rosin and turpentine will be turned over to him for sale on a two and one-half per cent commission basis.

The curve recording the prices of naval stores over the years resembles nothing so much as a design for a super-roller-coaster. The whole trade is as stable as a bowlful of jelly, for the brokers who sell rosin and turpentine here and abroad have little reason to want steady prices. With their inside knowledge, they frequently play both ends against the middle and profit most in the up-and-down swings of the highly speculative market.

Plump into the midst of this scattered, happy-go-lucky business jumped an eccentric inventor with a new kind of naval stores plant. It was not a still at all. It did not even work on crude gum. It was a big retort which he filled with chips cut from rich pine stumps. Homer T. Yaryan did not try to distill this mass of wood chips: he heated and softened them with live steam. Then he poured in a solvent which extracted the rosin and turpentine stored in the roots, and also another product, pine oil. He separated these three products in a modern fractionating still.

The owners of the little fire-stills laughed at this expensive, complicated contraption. It would never be a competitor of theirs—so they thought—for it cost too much. Besides it produced a low-grade, dark-colored rosin; a turpentine that had a strange, unpleasant smell; and pine oil which at that time nobody on earth wanted.

But during the past thirty years, that original wood-rosin plant of Yaryan's has grown into a really elaborate aggregation of autoclaves, tall fractionating columns, automatic valves and delicate instruments to control temperature, pressure, and the flow of liquids through a labyrinth of stainless-steel pipes. And today the modern wood-rosin plant turns out not three, but more than three-score products.

This young scientific rival, which they so scornfully derided, has forced the operators of the cheap little fire-stills to revolutionize their simple, individualistic enterprise or else get out of business. Ironically, the benevolent regulation of the Agricultural Adjustment Agency, which stepped in to save the turpentine farmer, is shoving the little operators out by compelling this ancient agricultural industry to become a complex ultramodern chemical enterprise.

In naval stores circles, Homer Yaryan has become a mythological character. He was that sort of man; a memorable personality, thinking originally, speaking in vivid phrases, acting independently. He made several fortunes—among his inventions was the sectional bookcase which was standard equipment in the den of the Gay Nineties—and he went bankrupt several times playing with ideas far ahead of their time. His propellent energy and real ability impressed everything that he touched. Alternately he smoked black cigars and chewed plug tobacco, which he discharged at any range up to twenty feet with the accuracy of a Marine sharpshooter. He swore like a truck driver, and when his first solvent extraction plant at Gulfport, Mississippi, came into production, he made a classic report to one of his business associates: "The blinking rosin pours out of the blankety-blank vat in a stream as thick as a woman's leg."

Yaryan had more than his share of luck, both good and bad. His process worked. But he could not sell his dark rosin or his smelly turpentine, and no one had any uses for his pine oil. Along came World War I to solve his selling troubles because war shortages cured buyers of their fussy habits. After the war this colorful pioneer doubtless considered himself exceedingly lucky to find a buyer for his entire naval stores enterprise.

The buyer of the Yaryan plants and patents was the Hercules Powder Company. Knowing that the management was scouting chemical fields for an expansion, a wide-awake Hercules salesman had suggested this purchase. He knew Yaryan because he sold him dynamite for blowing up pine stumps. During World War I speed-up cutting of the Southern pine forests and reckless chipping raised a scare that vanishing pine trees would mean death to the naval stores industry; no living trees, no gum. Congress had investigated and published such a doleful report that to the alert Hercules executives rosin and turpentine from dead stumpwood appeared almost a commercial sure-thing.

About the time that Yaryan had started, a chemically minded family from Milwaukee became interested in wood naval stores. Hugo Schlesinger had a private venture making sizing out of rosin for paper mills, and in 1912, to assure his chief raw material, he launched the Newport Rosin & Turpentine Company at Bay Minette, Alabama. This was so successful that a second, larger plant was built in 1916, at Pensacola.

Neither of the two strong companies, Hercules and Newport, that had become seriously interested in naval stores from stumpwood nursed any illusions about their products. They knew their rosin was off-color and that their turpentine had a disagreeable odor. They believed, however, that these faults could be corrected. They expected confidently to find industrial uses for their unwanted pine oil. But they had no inkling that the whole naval stores market would go back on them.

Contrary to all pessimistic prophecies, the gum naval stores industry did not disappear. Cutover lands reseed so readily and the pines grow so rapidly in the favorable Southern climate that it soon became clear that if chipping is not reckless, if the fire hazard is decently controlled, and if some attention is paid to reforestation, a supply of crude gum from living trees is virtually inexhaustible. As a result, the high wartime prices of 1915-18, which everybody had expected to go higher, began to drop. Hercules and Newport faced the double problem of raising the quality of their products and lowering their costs. They simply had to find out why their products were inferior to those of the gum industry. This meant a searching chemical study.

To most of us, rosin is a hard, clear, amber-colored substance, rubbed on violin bows, dusted on wrestlers' hands. But tons of rosin go into our paper, soap, varnish, and linoleum, and, believe it or not, rosin is an acid, abietic acid, the cheapest of all the complex organic acids. Turpentine is a cheap, colorless thinner for paints and a useful solvent for fats and greases. But to a chemist turpentine is a mixture of organic chemicals, chiefly pinene.

Rosin and turpentine are one thing; abietic acid and pinene are something quite different. They are chemical raw materials, convenient packages of atoms that can be juggled into such valuable products as plastics, synthetic camphor, perfume materials, even rubber. So every time the price of rosin and turpentine dropped, the Hercules and Newport chemists have just killed two birds with one stone by turning these natural raw materials into other, more valuable products. Thus the companies got out of competition in a falling, glutted market for naval stores and fared forth into new chemical fields where profits were more attractive.

Even before the end of World War I camphor, which the Japanese had monopolized, and terpineol, for which perfumers would pay a handsome price because of its fine lilaclike odor, had both been synthesized out of turpentine. Today terpineol is made in such quantities, and so cheaply, that it is being used in making soaps, inks, paints, and bug killers. As for camphor, we completely freed ourselves of that natural monopoly. Although we must have five million pounds a year for such plastics as Celluloid, we were not caught short during World War II when Japanese supplies were completely cut off.

These two were only the start. Now the wood-rosin plants turn out scores of the most surprising products: a sweet licorice flavor much used in candy and chewing tobacco; a liquid that dissolves Bakelite; a powder, one teaspoonful of which added to a bag of cement makes it flow smoothly and set much more quickly. In the big meat-packing houses, Mr. Porker is dipped into a vat of melted plasticlike rosin and comes out in a glove-fitting suit that peels off, bringing with it the last, tiniest bristle from the tip of his snout to the end of his curly tail. Hercules has synthesized from terpenes a potent fly killer, Thanite. Newport has worked out the first commercial process for making isoprene, the chemical mother of natural rubber, and a most valuable ingredient for compounding with synthetic rubber. Several pretty war babies have been born in the laboratories of both companies. These likely youngsters will go places and do things in the postwar world.

Wood-rosin men are modest about all this sensational chemical juggling. What they like to boast about is that their big plants, their laboratories, their array of new chemical products, are all built on the salvage of a waste—dead pine stumps, relics of the lumbering days.

If they pose as public benefactors, it is because their bulldozers are clearing over four hundred thousand acres of cutover land a year. That's a lot of fresh land for crops and grazing.

From the pine stumps yanked out of these cleared acres, the wood-rosin plants are now extracting more than half of all the rosin and turpentine produced in this country. The two pioneers have been joined by four smaller, tiptoe-alert companies—Mackie, Continental, Phoenix & Dixie, and Crosby—and Yaryan's moon-eyed notion of wringing profits out of dead stumps has become a real industry.

Long since, wood-rosin operators have learned to make a rosin as pale and a turpentine as nearly odorless as the finest grades made in the fire-stills out of gum from the living trees. From an arrant nuisance, their additional product, pine oil, has become a cheerful money-maker. In 1932, when I first visited the Hercules plant at Brunswick, Georgia, pine oil was oozing out of storage tanks faster than they could be built, and in the laboratories the chemists were scratching their pates baldheaded trying to think up uses of this oily flood. In 1945 at this same plant, I found them deliberately taking good turpentine and making it into pine oil to fill imperative orders for essential munitions needs.

But that is not all the incredible story. During World War II so active was the demand for certain critical naval stores chemicals that the wood-rosin plants eagerly took in gallons of low-grade turpentine recovered from the pine-wood chips cooked to make pulp in the Southern paper mills. They were even forced to go out into the market and pay ceiling prices for gum turpentine. The wood tail is not wagging the gum dog, it has him by the nape of the neck, shaking as a terrier thrashes a rat to and fro.

Long since, the owners of the gum stills have ceased to snicker at the elephantine wood chemical factories. Scattered and disorganized from its very beginning, their own business became more and more obsolete, and there seemed to be just nothing they could do about it. How could they join forces to put up a united front against this scientific giant? What could possibly be done to standardize the crude gum they took in from scores of turps farmers? Where was the money coming from to hire the technical brains to modernize their plants and their products? Year after year the going got tougher and tougher till suddenly the Great Depression came to block the way with a yawning chasm into which plunged the dwindling demand for their products.

If the little fire-stills disappeared, what would become of the turpentine farmers? The crude gum they collected from chipped trees was as valueless as a sand pile in Sahara, if there were no stills to convert it into rosin and turpentine. Fifteen hundred gum producers—

a few big operators of large gum rights, but hundreds of little rugged individualists who sorely needed gum money to eke out the scanty cash incomes of their farms —were threatened with commercial extermination. Some twenty thousand Negro chippers and dippers would swell the Southern relief rolls, and more seriously, would create a new, permanent Southern unemployment problem. It was no comfort to anyone that they were all, still owners and gum producers and all their Negro workers, the victims of scientific progress. Technological unemployment buys no more slab-side pork and cornmeal in Georgia and Alabama than it does potatoes and flour in Detroit and Seattle.

The Government came to the rescue. In 1936 Congress declared by law that rosin and turpentine from gum are agricultural crops and that rosin and turpentine from stumpwood are industrial commodities. This was done to assure the little gum people a share in all the generous farm benefits and to make sure that the half a dozen companies, big and little, in the wood-rosin industry got none. For this purpose of the Agricultural Adjustment Agency this charming fiction was quite understandable. When it was adopted by the Commodity Credit Corporation as a basis for ceiling prices, it became illogical and discriminatory.

The naval stores market was overloaded, so the Agricultural Adjustment Agency aimed at crop reduction by paying a bonus of three cents a cut face for not working trees under nine inches in diameter. Jay Ward, in charge of this conservation program, reported last year that $9,000,000 had been paid out. That averages $6,000 to each of the South's fifteen hundred gum producers. No wonder that the turpentine farmers clamored insistently for continued conservation benefits, though they knew very well—if they but stopped to think—that restricted output and high ceiling prices would further encourage the use of substitutes for turpentine and rosin in the paint, varnish, paper, soap, linoleum, and other industries.

In this respect the gum farmers were no more wise, no less selfish, than the cotton planters, or for that matter any organized group which during depression and war has taken advantage of abnormal economic conditions to grasp immediate benefits or to worm themselves into temporary positions. From the long-haul point of view, to get while the getting is good is seldom sound policy, and yet the conservation payments have had certain effects of permanent value. They have encouraged more careful chipping and cleaner collecting, and given the gum farmer ready cash, enabling him to get out of debt and putting him in a position to meet the revolution through which he is now swirling.

The other part of the Government relief program, loans from the Commodity Credit Corporation, were handled through the American Turpentine Farmers'

Association Cooperative, whose members pay assessment dues of ten cents a barrel on their crude gum. So long as market price was below loan value, the Commodity Credit Corporation continued to accumulate stocks until it had one million five hundred thousand barrels of rosin. The war boosted the price, baling out the Commodity Credit Corporation and cutting down the membership in the turpentine cooperative by half. This, too, is right on the beam of human nature, but again it waves the red flag, warning how dangerous it is to try by legislation to revise the law of supply and demand.

Wartime price controls, frankly adopted to benefit the gum farmers even at the expense of the wood industry, have backfired. The attempt to maintain an artificially high price for gum turpentine has resulted in enhancing the chemical values of wood naval stores.

For a century turpentine has been the number one product of the naval stores industry. The chief use has long been as a solvent in the paint industry, but for several years turpentine has been losing this market to mineral spirits. These solvents prepared from petroleum have steadily improved in quality and lowered in price. Gradually they have replaced turpentine as an essential ingredient in paint manufacture, banishing it to the secondary market as a thinner used by painters and householders. As J. H. McCormack, Newport's veteran president, put it, "Twenty years ago, we sold turpentine by the tankcar to the paint industry; today, we will sell it in five-gallon cans through the retail trade."

The turpentine that now goes into paint is but a drop in the bucket compared with what is being used in making the new chemical products. War needs greatly increased the demand for these chemicals, but the wartime ceiling price of eighty cents a gallon failed to recognize this switch in use. It utterly ignored the chemical values in turpentine as a raw material for chemical synthesis in competition with other solvents and chemicals. To make matters more complicated, the Office of Price Administration seriously underpriced rosin at five cents a pound.

Any question of future prices calls forth a lot of hemming and hawing, but the chemists figure that on the cold-blooded basis of chemical values the new ratio should be turpentine twenty cents, not more than thirty, and rosin, ten cents. From eight and five to twenty and ten will call for some pretty violent postwar readjustments.

As relief, the Government conservation-loan and price control program served splendidly; as a remedy for the deep-seated ills of the gum naval stores industry it was about as useful as a teaspoonful of paregoric in the treatment of bubonic plague. Yet the right cure has long been recognized.

During the past fifty years bold adventurers have often tried to establish large, centralized gum-processing plants. They dreamed of operations big enough to justify the cost of modern apparatus, profitable enough to pay a staff of experts. In such plants high-grade, standardized rosin and turpentine could be produced and at least the simpler chemicals of their wood-rosin competitors might be duplicated. During the six months of the gum-collecting season these visionaries worked their larger, more costly plants day and night only to land in the same boat with the owners of the cheap little fire-stills. Both were left holding a full year's stocks of finished rosin and turpentine. Unfortunately the ambitious ones held larger stocks and their more delicate, more elaborate stills deteriorated faster and ate up more carrying charges. Usually these dreamers were awakened by the sheriff nailing an auction notice on their plant door.

In spite of this long record of failure, whoever thought the naval stores problem through could see that the fire-still was as obsolete as the stagecoach and that unless the new chemical values could be extracted from gum, hundreds of gum farmers were simply out of business. Only the centralized plant to process gum in the modern style could save the oldest American industry.

At least, that is how the officers of the Glidden Company figured it out when, in 1935 at the zero hour of naval stores distress, they determined to take one more try at that oft-tried, oft-failed idea. This time it has gone across because three men as different as A, M, and Z, put their heads together and evolved a couple of brand-new, extremely simple ideas.

In their little shack of an office tucked away in the big plant near Jacksonville, these three told me their true fairy tale. Like the three genii in the *Arabian Nights,* each of this trio has distinctive gifts. E. W. Colledge, the general manager, is a wise old naval stores merchant who knows that tricky game as only years of rough-and-tumble play can teach it. McGarvey Cline, who was trained at the famous Forest Products Laboratory, has uncanny skill in handling sticky gum, complicated apparatus, and human beings. Joseph Paul Bain, Ph.D. from the University of Florida, is a silent scientist, well versed in the new chemical techniques.

"The plausible idea of centralized processing plants always bogged down," began Colledge, "because, like the fire-stills, they operated on a seasonal basis. Our idea was to take in crude gum, store it, and process throughout the year to fill current orders. If we could do this, we would at one stroke smash the bottleneck and cut out expensive financing by creating a spot market when collectors could sell their crude gum for cash."

"We did just that," interjected McGarvey Cline. "Gum sinks in water and in those tanks"—he pointed out the window to a row of big, shiny, aluminum-painted tanks—"under a blanket of water, it is kept safe from any chemical or physical changes. Simple, isn't it? The wonder is, nobody ever thought of it before."

"But we did have another good idea," continued Colledge. "We had to devise some method of classifying crude gum on the spot in grades acceptable to both the collectors as sellers and to ourselves as buyers. When brought in to us now, gum is divided in four grades and dumped into the proper one of four big mixing vats. This rough but satisfactory selection into four lots and mixing large quantities to average out quality is one of the real tricks of our trade."

Later we went out into the plant. The gum was pouring in at the beginning of the season, and I saw those four lots of thick, gray syrup, full of twigs and bark and pine needles, being scooped up in a half-ton clamshell bucket, swung overhead, and dumped into the storage tanks.

"We had some lucky breaks," admitted Colledge. "In a booming market we might never have gotten enough gum to start. The wood-rosin people also helped us by example and their accumulated knowledge. Last, we were helped greatly by the Commodity Credit Corporation, which was persuaded to finance the crop by paying the collectors for their gum as we took it in. We guaranteed Commodity Credit Corporation against loss through improper manufacturing or grading, and the fact that these loans have been repaid proves that crude gum in a central store can be a commercial commodity, a safe security, and a quick asset."

Today the Glidden plant operates the year 'round and so successfully that a duplicate has been built at Valdosta, Georgia. At Jacksonville seventy-five thousand good dollars have been invested in a ninety-eight-foot stainless steel column fitted with the latest control gadgets in which Dr. Bain is making exciting experiments in the chemical possibilities of gum products.

Others have followed these leaders. Having no desire to be left out of the new picture, several old, established factors have put up the money to build centralized processing units and more and more of them are cultivating the chemical side of the business. A newcomer, the Filtered Rosin Products Company, for example, has five of its own modern stills at strategic points in Georgia whence they ship pure rosin and turpentine to the supercentralized chemical plant at Brunswick. Here a startling new synthetic shellac, christened Chemlac, has been perfected in cooperation with Corn Products; a logical alliance since Chemlac is made from rosin and zein, a by-product of corn starch.

The hurricane of progress that is uprooting the old uses of naval stores will leave the piney woods of the South dotted with deserted fire-stills, picturesque relics of a romantic chapter in our industrial past. But these stormy changes are not driving the gum collectors into cyclone cellars—quite the reverse, they are stepping forth briskly, toting their dripping, sticky barrels to the new gum-chemical plants and carrying home the cash.

Thanks to federal aid the turpentine farmers have achieved an unaccustomed financial stability, freedom of thought, and independence of action. All this is good, but better, they have learned wholesome lessons in up-to-date methods applied to their rule-of-thumb craft. Only the incorrigible old-timers persist in the destructive habit of deep-chipping the trees. Hundreds of the progressives are trying out the new "plumbing system," a very light cut through the bark only, followed by a spray of sulfuric acid which stimulates the flow of gum. The yield per tree is thus doubled. Down at the Southern Forest Experiment Station at Lake City, Florida, unorthodox cuts and strange chemicals are being tried out by Harold L. Mitchell to make the trees give up more gum. He is even carrying on plant-breeding experiments to find more prolific strains of trees.

To the turpentine farmer of yesteryear this last would seem to be "tops" in ridiculous nonsense. Today throughout the Naval Stores Belt, the results of Mitchell's scientific work are as eagerly awaited as the first box score of the World Series. And that little bow to science is the most revolutionary change of all in the remaking of naval stores into an ultramodern chemical business.

8

Pine Inventory

IN THE PINE WOODS of the South the silence is punctured and ripped by new noises, the staccato put-put-put of a laboring gasoline engine and the snarly whine of a circular saw chewing through soft wood. Like trumpets of the heralds proclaiming a new royal decree throughout the length and breadth of some ancient domain, these strange clarion calls announce a new régime: machines have gone to work in the woods. Southern lumbering operations are being rapidly mechanized, and whenever man applies power to any job, important and often quite unforeseen things are bound to happen.

The war created an acute labor shortage throughout the pine woods from southern Virginia to East Texas. Being primarily a labor-saving device, the machine was brought in to keep up the supply of lumber and pulpwood that was essential to the war effort. This compulsory innovation will have effects upon the South that reach far beyond the sawmills and the paper plants.

Getting out the wood is rough, rugged work. In the lumber camps the draft boards found few physical rejects and they made no occupational exemptions. What they started, the shipyards finished. Swinging the shipwright's mallet is child's play to holding up your end on a crosscut saw, and the rip-roaring tradition of Paul Bunyan was no charm whatever against pay of fifty-seven cents an hour instead of the minimum legal wage of forty cents.

Looking back on the war experience, it seems a wonder that any men stuck to lumbering although wages doubled and sometimes trebled. For getting out as much wood as possible, these high wages were a questionable help. Whoever knows the Southern Negro knows that when he is paid a couple of dollars or more a day, he simply quits work after he has earned his modest needs of five or six dollars a week. In these materialistic days this easygoing philosophy has a good deal to commend it, but not to a woodlands superintendent hounded to make a quota so that the mill can keep up the schedule of imperative Army and Navy orders.

The Southern paper mills were caught tight in a bad jam. On the one side were more orders than ever before for kraft paper that must be sold at a ceiling price no matter how costs rose and time pressed. On the other side less labor was being paid higher wages.

Relief by means of machines that would cut and trim more logs, load and unload them more quickly, was the obvious solution. Accordingly, a number of different paper mill managements began mechanization soon after we entered the war.

There was one man, however, who had thought about cutting and handling machines and even tried them out before Pearl Harbor. Edward J. Gayner, III, of the Brunswick Pulp and Paper Company, deserves the credit due a pioneer. A businesslike man, businesslike in thought, word, and deed, he studied the whys and wherefores of this new development, and in orderly fashion figured out that mechanization of the woods operations was sane and sensible.

"Most of the paper mills that came South during the past twenty years," he told me, "were attracted by wood and labor, both abundant and low-cost. Wood was cut and trimmed, loaded and unloaded by hand, arduous jobs all of them. There was no doubt about the low cost of stumpage; that is, what we paid for wood cut on another man's land. But I began to wonder if the cost of getting it into the plant was really as low as we liked to think. Compared with Northern or Western scales, our wages were certainly not high, and yet right out my office window I had a constant object lesson over in the woodyard that set me to figuring. We tried out a couple of crane-and-chain loading experiments. At that time nobody was very much interested. Now we all wish we had mechanized a lot more a lot earlier.

"Our old crew for wood handling at the mill was two hundred and ten men. Now we handle fifteen per cent more wood with one hundred and thirty-three men. Regardless of costs, we simply could not have kept going during the war without that saving in men. But the saving is real, and it is even greater in cutting and loading out in the woods."

Machines for handling the unwieldy logs were there ready-made, waiting to be adapted to the backbreaking work of loading and unloading. The need for machines to replace the ax and cross-saw was fortunately met by a burst of invention that perfected several types of portable circular and chain saws just at the right time.

In January 1945, when I stepped into the tiny office of Victor Schoffelmayer, his familiar bright eyes and pointed nose peered at me over a stack of letters piled high on his big desk. They were all inquiries in response to the picture of a new portable circular saw which he had published in his agricultural column in the Dallas *News*.

"Just look at 'em," he exclaimed, screwing his cheerful countenance into a semblance of dismay, "I never had such a flood of correspondence on any one article

I ever wrote." Then he smiled all over. "It's great, isn't it?" he exclaimed, patting the stack of letters affectionately. "They have come from the Rio Grande Valley to North Carolina, from big lumber companies and little farmers and scores of them from hardware and implement dealers. Even when you know you are breaking a big story, it is not often that you get anything like that kind of immediate response. And you believe me," he went on excitedly, thrusting in my hands a glossy newspaper photograph, "that overgrown power lawnmower with long handles and a buzz saw stuck out in front is the most revolutionary machine introduced into the South since Eli Whitney's cotton gin."

Victor is an enthusiast, but he is shrewd and for twenty years he has watched Southern agriculture with the keen eye of a newshawk. Clearly he foresees what an inexpensive, one-man machine to fell and trim trees means to Southern farmers and to big sectors of Southern industry. He confesses ruefully that he wishes he might guess what it will do to Southern wages and working conditions.

A good case can be made for the Southern pine as the most important element in the revolution that is stirring throughout Southern farms and factories. Certainly it touches many more Southerners than do gas and oil. It hooks in closely to the grave cotton problem in the very sections where it is most difficult to solve. Wood, which is man's oldest structural material, had seemed on the way out. But wood has of late enjoyed a revival as enthusiastic as an old-fashioned camp meeting.

By new chemical treatments wood is now rendered flexible, fireproof, and amazingly strong and durable. Paper-thin sheets of wood smeared lightly with a heat-setting plastic can be built up, layer after layer, over a form. Thrust the whole into a pressure chamber; clamp down the doors; turn in live steam up to a pressure of hundreds of pounds per square inch. Within a very few moments you can withdraw a canoe, a bathtub, the body for a baby carriage or an automobile, a housing for a carpet sweeper or for a huge dynamo. The plastic plywood has been formed hard, strong, without flaw or seam. I have lifted with one hand such a bathtub. It is coated without and within with a luscious, enamellike finish that will not chip and cannot stain; waterproof and acid-and-alkali-proof; capable of being colored permanently any shade from jet black to snow white.

This fashioning of plastic plywood over forms seems so simple and its possibilities are so bright, one is likely to overlook the fact that here is an absolutely new technique to handle a new kind of material. For ages man has pounded and molded metals, chipped stone, sawed and shaved wood, woven fibers, and then fastened these various materials together by weld, by nail, by glue, by thread. But here we have a new fabricating technique that has no past in the handicrafts. It has a great future in the coming age of chemical technology.

With new plywoods and lumber improved—so as to give new properties and greater strength and durability, wood will also be important in the South for its new chemical values. It makes a vast difference whether one thinks of naval stores as rosin and turpentine or as abietic acid and pinene. So wood as wood is a far different material from wood as lignocellulose, which embraces all the vegetable kingdom—oak and pine, apple and mahogany, bamboo, cornstalks, bagasse, even straw. From lignocellulose come rosin and turpentine, the great natural chemical monopolies of the South, and cellulose, the raw material of paper, cardboard, and rayon, the great rivals of the South's cotton. From lignocellulose we also obtain the tannins. Furthermore oils, fats, and the so-called wood sugars make up a fifth of all the woody substance, and finally lignin, a third of all wood tissues, which has been hailed "the greatest waste in all American industry." We shall soon turn this lignin waste into wealth.

In the startling renaissance of wood, the pine forests, despite reckless cutting, abuse and neglect, careless cattle-grazing and consuming fires, are still one of the South's dominating natural resources. The South is the great agricultural region of this country, yet sixty per cent of all Southern land is timberland; more than two hundred million acres of forests, more woodland acres than in 1860. Throughout the South are scattered seventeen thousand sawmills, most of them little "grasshopper" mills that move from place to place. Every year they stack up forty per cent of all the lumber cut in the United States. From these same Southern forests comes the pulpwood—six to twelve inches and under, too small to make lumber—to make eighty per cent of all American kraft paper.

This enormous treasure house of lignocellulose is not pre-empted by big interests. Three-quarters of the landowners hold fewer than five hundred acres each. All the paper companies together own only five million acres: big lumber operators but twice that much. Between them they control less than a tenth of the total acreage. Within this vast area of small, individual ownership, over half a million workers are employed in the woods and the sawmills.

Yet Southern woodlands are still resources that are not utilized to their fullest. The foresters who have cruised the territory making expert appraisals tell us that these woodlands could grow at least twice as many trees. By intelligent care they could provide an adequate, year-after-year supply of pine for lumber and pulpwood and naval stores, a perpetual inventory of raw material giving regular employment to a million Americans.

But it is no longer possible to double the value of

Southern wood products by the simple expediency of doubling the lumber crews. Production of pulpwood differs radically from the production of pulp. While the output of pulp may be doubled by installing twice as many digestors and doubling the purchases of chemicals, nevertheless that plant inexorably depreciates to zero value. On the other hand, the living tree reproduces itself naturally. Even the best-cleared farm land, left to itself, in time returns to forest. These plain, uncompromising facts make a Southern woodland more important than the gum turpentine or logs it currently produces.

Anyone broadly viewing the future opportunities of the South recognizes this instantly. However, many landowners, big and little, are so nearsighted that literally they cannot see the forest for the trees. Though the broad theory is generally accepted, its application is likely to be narrowly selfish.

Eating the seed corn upon which depends next year's harvest has been a bad habit of *Homo sapiens* ever since the caveman planted his first crop. But careless profligacy puts a high premium upon security. Let the thoughtless landowner once get the idea of his woodland as a savings bank, then he is likely to become an enthusiastic conservationist. Nowadays his good resolutions are receiving constant encouragements.

In the first place, pines of several varieties grow so rapidly in the South that they hold forth promise of tangible returns within a reasonable time. To most men, planting trees for posterity makes a pretty weak appeal to the pocketbook motive, but here is a tree crop a man can cultivate with the prospect of harvesting it annually himself. In addition, if handled correctly, he can leave a valuable legacy to his children, which comes close to eating one's cake and having it.

An inch a year added to the diameter of a growing pine sounds too good to be true, but I have measured it. Mounted on the office walls of a lumber company at Crossett, Arkansas, and of a paper mill just outside Mobile, Alabama, are the planed and polished cross sections of pine tree trunks that measure—I laid a ruler across them—eleven inches. The clear, dark, annual growth-rings—I counted them—number exactly ten. A hard-headed lumberman in Jacksonville told me of young pines that made up to twelve inches diameter, two feet above the ground, growing between the ties of the roadbed of a departed railroad. "It isn't ideal soil," he commented dryly, "but it kept down competition, and the end of railway traffic gave a date fixed beyond any argument."

Those trophies on the office walls are in reality targets. Admittedly, they are specimens above the average growth. They are not so exceptional, however, but that with care and climate they may not become the average. There is plenty of latitude in climate: the entire coastal plain from North Carolina to East Texas. There is not perfect agreement as to the best system of care. It is admittedly true, however, that any kind of forestry practice means quicker growth in the long run and quite promptly an appreciable year-to-year income. Pine-tree cropping in the South very quickly comes to yield an annual harvest.

The pines deliver three different cash crops, and a turpentine farmer, a paper-mill superintendent, or a lumber dealer will each tell you his own very definite ideas. Each interest is distinct, sometimes conflicting, but fortunately never incompatible. Oil and water can be mixed by means of an emulsifying agent, and in this case the best interests of the woodland itself, which is also the best interests of the owners of most of these acres, is not to produce any one of these specialized crops, but all three.

Pines growing on the land of a paper-making company that was thoroughly shortsighted, might never be allowed to grow beyond the eight- to ten-inch diameter which is the ideal size for pulping. Nevertheless, wise administration of that woodland would leave seed trees amid even the most conscientious cutting for pulp purposes. In fifteen or twenty years, these seed trees would be ready for lumber.

The "cat-faces" of the turpentine farmer are popular neither in the pulp mill nor the sawmill. If the turpentine chipper cuts so that he kills the tree or weakens the stem so that it snaps in the next hurricane, then he becomes a public enemy of the woods. Both of these tragedies are much more likely to happen if a very young tree, under good pulping size, is tapped. This itself is a forestry crime.

The lumberman no longer has the opportunity to go into a virgin stand of pines and cut clean as he did in Louisiana and Texas. In this region where once stood the finest yellow pine forests on the continent, there now stretches mile after mile of terrifying object lessons: millions of acres stripped so bare by ruthless cutting and repeated fires that they can be brought back to forestry only by expensive replanting. Throughout the Southern states there are eighty million such devastated acres. But that type of utter destruction is fast passing.

It is a healthy sign that war conditions—high prices and scarce, highly paid labor—have stirred up a conflict of ideas about proper cutting. Nobody any longer argues the ideal system, but farmers and county agents on the one side, and on the other, wood bosses and foresters, accuse each other of wasteful practices.

Ask almost any farmer and most of the county agents, and they will tell you that the paper people are no longer to be trusted. The pressure to get out all the pulpwood possible is so great that "they'll strip off everything from three inches to a foot."

Repeat this accusation to a paper-mill man and he will reply, "No doubt a lot of landowners feel that way about us. No doubt, too, the high price of stumpage and the necessity of producing paper under ceiling prices are a big temptation to cut clean in order to cut costs. After all, the cost of wood includes not only the wood itself, but cutting and hauling. Wages have doubled, while the haul has increased from five to, say, twenty miles. But high prices tempt landowners, too. We try to cut strictly according to their directions, and many, many times they tell us to cut clean so as to clear the land for cattle range. Sometimes that's just an excuse, and the cattle never come in."

Nowhere is the God-given advantage of Southern climate more richly demonstrated than in the pine woods. The long growing season and the generous rainfall of the coastal plain make not only for exceedingly rapid growth but also for quick and easy reseeding. This is very important, since replanting is costly and risky; at least five dollars an acre, and at best, only sixty per cent survival of the baby trees with always the chance of destruction by even a light fire. Only in sections where the land has been stripped naked by ax and flame is replanting necessary. In such areas, both the lumber and paper companies are doing more and more of this work and encouraging the farmers in the neighborhood to follow their provident example. Ten million seedlings a year are being distributed in this way. Each tiny plant is an eloquent propagandist for better forestry. Any man who sets out a thousand seedlings, even if he hires someone to do this backbreaking, fussy job, will see to it that those tiny tufts of green needles are not burned to death. He has acquired at once a dollars-and-cents stake and a forward-looking interest.

Pine trees do not bear seeds every year. There is a good deal of uncertainty, depending upon growing conditions and considerable variation in the different species. The long-leaf pine, for example, fruits only once in seven years; the short-leaf pine, every two or three years. For any variety, however, a couple of big, thrifty parents to the acre are enough to assure the oncoming generation.

Cultivation helps and it also pays. This does not mean tilling the soil and applying expensive fertilizers. Roughly plowed strips of bare soil, six or eight furrows wide along the highways, stop many an incipient fire from a careless cigarette. Such "fire strips," zigzagged through the woods, check blazes that might sweep over wide areas. In the ordinary sense, this is all the cultivation needed to grow a timber crop, and "crop cutting" is in truth a form of the simplest kind of cultivation: protection plus intelligent thinning to give the trees the best chance for development.

Crop cutting is quite different from selective cutting. The latter is cutting off the timber but leaving the seed trees. The first involves careful cutting almost every year to keep down brush and take out the crooked and diseased trees. Most of these can be used as pulpwood. Later the pulpwood proper is cut with a wise eye to spacing. Next the turpentine trees are scrupulously selected and chipped. Finally, there is a harvest of poles, piling, and lumber trees. And by that time the younger growth has come along, ready to repeat the cycle.

Under this system it takes fortitude to permit the trees to grow to timber size. However, since they increase in girth around the outside, the amount of wood produced multiplies greatly year after year. Since the price of sawwood promises to advance proportionately more than pulpwood, it becomes an increasingly wise gamble. If the landowner is to gather his third crop and go in for turpentine, he will be wise not to chip faces on trees less than eight or nine inches in diameter, and then to continue this cutting for at least three or four seasons. Even the modern method of light bark-chipping does definitely retard the growth of the tree, and the chipped faces cut down its value either as pulpwood or lumber. The trees should therefore be compelled to make good these losses by producing several crops of gum.

To burn or not to burn?—that is the question. Sincere advocates of controlled burning to keep down the undergrowth are by no means confined to the backwoods cracker who follows grandpappy's practice of burning the woods each spring to encourage the growth of grass for his scrub cattle. This type of burning, according to the soil conservationists, is utterly without justification. The young grass that springs up so fresh and green against the charred background creates but an illusion of lush fertility. As a matter of fact, the annual burning not only prevents reseeding, since the tiny seedlings are cremated in their babyhood, but it also destroys layer after layer of precious humus in the soil, leaving it dead and infertile. Education is the only cure for this senseless practice. Yet there are well-educated men, even expert foresters and experienced woodsmen, who advocate controlling burning. They maintain that this should be promptly done after selective cutting, in order to keep down the first luxuriant growth of underbrush and present the most favorable opportunity of reseeding. They stoutly present a parade of argument that heavy underbrush increases the fire menace to the point of uncontrollable blaze. Naturally these beliefs are most firmly held in Florida and along the Gulf Coast where the spiny palmetto springs up luxuriantly in cutover land.

Countering these persuasive statements, the nonburners point out that a controlled fire may easily become uncontrolled. Furthermore, they claim that in keeping down the underbrush, several crops of promising seedlings must be slaughtered, and they question whether the young trees either start or grow as well on naked, fresh-burned land.

All schools of thought vote that at heart the problem is one of education. Within the past five years there has certainly been an appreciable decline in the fire hazard. In this work the big companies are doing their full share. Among the paper mills it is virtually standard practice to acquire more and more land and to practice scrupulous forestry. Even the lumber companies, who of necessity must look ahead more years for profitable returns and who have greater temptations to cut clean, are increasingly interested in the perpetual inventory idea. Among outstanding examples of this long-haul point of view are the Crossett interests in Arkansas, Crosby in Louisiana, Sullivan and Putnam in Florida, Smith in Alabama, all held up as shining examples of good practice. Rural schools and agricultural county agents are preaching the gospel. Alabama has recently proposed the most progressive measure of all: a state severance tax on turpentine, pulpwood, and lumber—all the pine forest products—estimated to yield a million dollars a year, and to be spent in fire prevention.

9

Pulp, Paper, and By-products

IN MICHIGAN one crystal autumn day, two chemists were hunting ruffed grouse when they flushed an idea as startling as the whirring rush of that king of Northern game birds as he bursts in full flight from underfoot—pine trees could be a greater annual crop in the South than cotton.

Fifteen years ago you might just as well have said that Rhode Island will someday be larger than Texas or that we will be willing to pay more for sand than for sapphires. Tomorrow that forecast may be literally true.

These mighty hunters were "Capt'n Charlie" Herty and "Uncle Billy" Hale: Dr. Charles Holmes Herty, once professor of chemistry at North Carolina University, and Dr. William Jay Hale, director of organic research at the Dow Chemical Company. They lunched that day on a little open knoll in the midst of dark jack pines and flaming maples, all scarlet and gold. The October sun was warm, and after munching gratefully on their smoked sausage sandwiches, they leaned back against a moss-covered log and talked and talked; talked till the chilly evening shadows interrupted them. Both have indignantly denied as apocryphal and scandalous the story that most of that afternoon a cock grouse scratched ant eggs around the old stump against which they had propped their guns. But it is well known that the annual hunting parties of these two chemical cronies were more famous for the bag of live ideas than of dead birds.

Consummate organic chemists, both of them, they naturally talked organic chemistry—not the formal chemistry of carbon compounds, but the living chemistry of farm crops which manufacture out of air, water, and the minerals of the soil, starches, oils, sugars, proteins, cellulose. In terms of the miraculous efficiency of the living plant that, with the aid of solar energy, makes these amazingly complicated compounds out of the cheapest, most available raw materials, they pondered the age-old farm problem—then a blazing political issue—and they probed the unsound core of the philosophy of scarcity.

"Henry Wallace plowing under corn and cotton," said Billy Hale, "is exactly like the pet bear who killed a fly on his master's nose with a rock, which cracked his skull."

"This ascending spiral of higher prices and higher wages," generalized the thoughtful Herty, "ends inevitably in ruin. More goods sold more cheaply is the only way to raise living standards substantially and permanently. During the last century, we have registered greater materialistic progress than in all human history because the machine helped us to create a real abundance. That abundance already threatens our supplies of irreplaceable metals, gas, oil, and coal. In the coming century, we can continue that progress only by using our materials more intelligently."

From there on in, they appraised the pressing problems of those dark depression days in the illuminating light of chemical values. That afternoon the idea of chemurgy was born in Dr. Hale's fertile brain, and out of that discussion popped a question that deeply stirred Herty, both as a Southerner and a chemist.

Is it possible that cellulose from quick-growing Southern pine is cheaper than cellulose from cotton?

Six years afterwards, sitting before the blazing logs on my hearthstone, this patron saint of Southern research told me his own answer to this haunting question. In his little Savannah laboratory Herty's epochal experiments in pulping and bleaching young pines had come to a triumphant conclusion. At that time the price of cotton had been already pegged for seven years and our cotton exports had been cut in half. Even in 1937 this forward-looking Southerner saw quite clearly the role that cotton and pine—as cellulose—were to play in the work and wealth of the South during the strenuous years that we have just been living through, and he peered ahead prophetically into the years ahead of us.

"The world will not buy our cotton," he began, "if it can buy cheaper from Brazil or China or India. Rich as we are, the American people cannot afford indefinitely to tax themselves to maintain a position of economic isolation in one of the world's great commodities. Inevitably, our cotton acreage must be reduced, and surely the first acres to go will be those where costs are high and yields low; the worn-out fields of the coastal plain from North Carolina through to Mississippi.

"Here is the very country," he exclaimed, leaning forward eagerly, "where all species of Southern pine grow most readily and most quickly. To replace the cotton crop with a wood crop is the simplest solution.

"Those depleted acres seldom produce more than half a bale of cotton. At the rapid rate of growth of pine in this section, they can produce, year after year, half a cord of pulpwood. Half a bale of cotton equals, say, two hundred and fifty pounds of pure cellulose; half a cord of pulpwood, about five hundred pounds. Planting, chopping, and picking are a lot more work than fire protection and selective cutting. By creating a steady demand for an annual crop of pulpwood, the paper mills will automatically solve the cotton problem in the very sections where it is hardest to solve. That's why I want to see Southern pulp made into newsprint, as well as kraft, so as to increase and stabilize that demand."

That husky paper-making infant of Southern industry whose cradle Charlie Herty rocked, has become a giant. How thrilled he would be to know that since his death in 1938, this industry has doubled. Fifty mills throughout all the thirteen Southern states, except Kentucky and Oklahoma, are pulping four million cords of wood a year. Its investment now totals over two hundred million dollars. It gives employment to over forty thousand men. How proud Herty would be that of the two and a half million tons of kraft paper and board rolling off its gigantic machines, a fifth is bleached kraft, not a single ton of which was made when he began his bleaching experiments.

He never lived to see newsprint made from slash and loblolly in the pioneer mill at Lufkin, Texas. But he knew the inside of the fight against subtle, determined opposition from Northern paper and financial interests and the struggle to build that mill with Southern dollars. Today the Southern Newspapers Publishers' Association stands ready to back other Southern newsprint mills. Wood, water, and transportation facilities have all been surveyed for the best sites; the capital is ready when men and machines are again available. Four pages out of five of all Southern newspapers are still printed on paper made from Northern or imported pulp, and this loyal Georgia chemist would have keenly appreciated this bright sign of revived independence and purposeful self-help. How he would chuckle at the news that the demands of our global war forced the kraft companies to go out twenty, thirty, fifty miles in search of pulp. For this greater demand is a sharper spur to the cultivation of pulpwood as an annual crop, hastening the day he foresaw when abandoned cotton land in his own state would produce a ton of good cellulose to the acre each year. Most of all, Charlie Herty would be interested in purified cellulose made at Fernandina, near Jacksonville, for rayon, and the new uses being found for the wastes and by-products of all Southern paper mills.

Though the Southern paper industry has grown so lustily, there is no ceiling upon its future expansion—except an adequate, permanent supply of pulpwood. Third only to water and milk, we Americans consume more paper than any other commodity, and we promise to break the old prewar record of three hundred pounds a year apiece. If the South were to add only sufficient newsprint for Southern newspapers, several gigantic mills would need to be built and the region's paper output increased an eighth. As for the tough kraft wrapping paper and the light, strong kraft board, war shortages of cotton bags, jute, tin, plastics, wood have shoved them out into new markets.

These new positions will not be held without a merry battle. Paper bags and cartons are the cheapest and lightest shipping containers. These are two telling arguments. The paper bag is better than cotton or burlap because it is dustproof and siftproof. But it has no salvage value as a washrag or a seat cushion. So it is all along the line, each wrapper or container has its own good points as well as its own drawbacks. The public, always fickle and seldom very logical, will settle all these arguments in the end. Wishful thinking has no part in the calculations of Henry Carruth, the lean, rugged, wrinkled executive vice president of the Union Bag and Paper Company. He is certainly prepared to expect the unexpected.

"A lot of new uses have been discovered for paper," he said, "but it's a funny thing how many of them have been additive rather than substitutive. Take the packaging of cigarettes: back in the happy days when your Dad learned to know Sweet Caps, he bought ten for a nickel in a simple jute-type carton box. Then the makers added pictures of Anna Held in tights and the pitcher, John McGinty, in sideburns. That was only bait, and yet, don't forget packaging is half bait, anyway. Only the up-to-date bait is attractiveness, safety, cleanliness, convenience, and a thoroughly appropriate holder for the goods. Today, the cigarette package is a multiple container: a cheap bond paper, metal foil, a paper label, cellophane, ten packets in a cardboard container. What it will be tomorrow, only the advertising boys know. Bread was first sold unwrapped, then wrapped in wax paper. Eye appeal—"see what you buy"—brought in transparent wrappings, and later, so the good housewife could know the brand she was purchasing, the transparent sheets were loaded with titanium pigments, making them opaque so as to take printing. There are styles in packaging, just as unpredictable as the heroes of the bobby socks brigades.

"The growth and strengthening of the Southern paper industry, as we see it," he went on thoughtfully, "are coming through diversification. More and more mills are equipping to bleach, and bleached pulp will be more important. We are going to use more hardwood, and a lot more tissue paper will be made. Certainly expansion is coming in alpha pulp for chemical processing. Competition is going to be keen—some of it, no doubt, from imported Scandinavian pulp—and because every penny must count, all mills must pay more attention to recovering their by-products."

Looking ahead at future Southern industrial developments, some of Henry Carruth's points bear examination. Let's start with the use of hardwood. Much of the Southern second-growth timber is made up of hardwood. Half of these hardwoods—gums, cottonwood, poplar, maple, and bay—can be pulped. The gums especially, as recently proved at the Herty Memorial Laboratory in Savannah, pulp as readily as pine, take much less chlorine for bleaching, and produce somewhat larger yields with less steam and chemicals. Even such dense woods as oak and hickory have been successfully pulped by the kraft process. Carruth believes that to eke out supplies, kraft mills will work in greater percentages of hardwood. Flintcote is already using as much as seventy-five per cent in their roofing paper, and at a kraft mill in Alabama, where they jealously guard their secret know-hows, I was told they use up to thirty per cent of hardwood and get, they claim, a better paper.

More purified chemical cellulose pulp for rayon, lacquers, and transparent film is a prophecy sure to come true swiftly. These chemical users of pulp have long clung to a sort of preconceived prejudice that the only proper alpha cellulose was prepared from sulfite pulp. This notion has been shocked twice.

Nitrating cellulose to make smokeless powder is an exceedingly tricky operation. Pure materials and scrupulous control are essential or somebody gets hurt. Shortly after the Germans invaded Poland, the British Admiralty sent Louis Hibbs to this country. In his own right Hibbs is an expert, but to double-clinch his authority, he is a representative of Cross and Bevan, a research establishment where viscose rayon was born and which since 1881 has been "tops" in cellulose chemistry. Hibbs came here to buy kraft pulp to nitrate for the British Navy, whose specifications for explosives are supersensitive. He made no secret that sulfate kraft pulp was actually preferred to sulfite for this most exacting process.

The other shock came when Rayonier, successful makers of alpha cellulose by the sulfite process on the

Pacific Coast, built a big, sulfate cellulose plant at Fernandina, Florida. They did not jump across the continent blindfolded. They shipped tankcars of local water along with cords of pinewood from Florida to Washington for their pilot-plant trials, to make the assurance of their high quality alpha cellulose doubly sure. The product of that new Southern mill of theirs has gone into all types of cellulose products from millinery trimmings to photographic film.

Chemical cellulose from Southern sulfate pulp cannot be "too bad," and more of it is going to be made and sold in the coming years. Three, at least, of the big kraft mill managements have delved into this postwar prospect and worked out the details of necessary modifications in their operations to produce it. This takes a bit of doing—chemical adroitness and considerable stainless-steel apparatus—but there is talk of a Southern production of over a million tons of chemical cellulose in the near future. And that talk is not cocktail hour gossip.

Diversification of this sort within an established Southern industry is very healthy. But more newsprint and alpha pulp are not the only additional products that will be coming from the paper mills. The Southern paper industry is beginning to turn its own wastes into new wealth, and will bring to the market at an opportune time lignin, furfural, tall oil, and turpentine.

After the big, initial investment in plant, a paper company's current expenses go chiefly into taking out of wood all of its constituents, excepting cellulose. In thus turning lignocellulose into cellulose, the paper mill gets from its paper-making machines only half the raw material it started with. The other half is pure waste; not only a loss, but a disposal problem. More than half of this waste is lignin, which has thus achieved the unenviable position of the greatest waste in American industry today; some two million tons a year.

This mocking challenge to the ingenuity of our chemists has not been dodged, but it has not been very successfully met. Lignin itself is a baffling material, so complex that even after years of study the exact formula has never been determined. But the chemists do know what happens when lignin is treated with many different chemicals in a lot of different processes, and they have proposed a multitude of uses for it. These run the gamut from adhesives to water purifiers. Ten years ago, a brilliant research team at the Rothschild paper mills, headed by Allen Abrams, evolved a clever process for making the flavoring substance, vanillin, out of lignin. This is being done, but though the most widely used of all flavors—more of it goes into cigarettes than into ice cream and candy combined—vanillin manufacture bores but a tiny hole in the mountain of lignin waste.

One big wood industry has always used lignin. Masonite, the hard, smooth building board, is manufactured by a novel wood-working process. In this unique operation, wood chips are fed into a steel cylinder into which live steam is then introduced. The pressure is raised to a thousand pounds, held there for a second or two, and then a hydraulic valve at the bottom is quickly opened. The mass of chips literally explodes from the "Masonite gun," rent into a fibrous mass. This is washed and then formed on a paper-making machine into thick, wet mats, which are pressed hydraulically into the familiar boards. By changing the temperature or pressure within the gun, by drying the wet mats before pressing, by adding tempering oils, plastics, resins, or other ingredients, many modifications can be made of the finished product. Naturally the Masonite research staff has been experimenting along these lines.

"Masonite material as it comes from the gun," explained Rodger Dorland, the assistant director of these researches, "is a semiplastic. Sparked by the war need for semicircular reflectors for fluorescent lights, we have been doing a lot of work in preforming and have developed an all wood plastic which ought to be cheaper than the cheapest synthetic molding material."

A really low-cost plastic—cheap enough to compete with wood in window frames, moldings, baseboards, and other building materials—is still an unfulfilled dream. Accordingly, this new Masonite material, which makes greater use of lignin as a binder and plasticizer, ought not to have much difficulty elbowing its way into the already crowded field of plastics.

But lignin itself has won a toe hold in two other markets. It's an axiom in the laboratories that if any chemical product is available in quantity at a reasonable price, some smart chemist will find a good use for it. Several years ago, the West Virginia Pulp & Paper Company, which for a long time has taken special interest in lignin, concluded that it would be a long step in the right direction to make this waste into a uniform, thoroughly standardized product offered in commercial quantities. In due course, they sent out generous samples of brown, free-flowing powder, which they trade-named "Indulin." But not content to let their new product sell itself, Dr. Arthur Pollak, in charge of the development laboratories at the Charleston plant, went right ahead by the good, old cut-and-try method, to find all and any possible uses for it.

It happened that at the moment rubber compounding was a topic of lively research. All the compounding techniques of natural rubber had to be revamped to suit the new synthetic rubbers. Carbon black, one big secret of the long life of the prewar tire, had become scarce and costly. Pollak found it was easy to enlist the cooperation of rubber company chemists in hundreds of experiments in which lignin was substituted for carbon black in proportions ranging all the way from one to one hundred. The results astonished everyone. Not only was the substitution successful, but indeed lignin appears actually to improve the compounded rubber, especially synthetic rubber.

From New England comes word of a use for lignin that would be almost without limit. Robert S. Aries at Yale, who has been poking about the forests and woodlots hunting chemical opportunities, has announced that lignin increases the fertility of depleted soil twenty per cent by adding organic matter and encouraging the formation of humus. Worn-out farm lands in every section of the country often suffer more from lack of living organic material than from lack of the elements, potash, phosphorus, and nitrogen. To revivify these "dead soils," as Louis Bromfield calls them, is the crucial step in soil building. If lignin can help, a notable "wealth from waste" triumph will be scored.

Wood seems such a solid substance, it is hard to believe that a fifth of the pulpwood that goes into the paper mills dissolves away in water and literally goes down the sewer. Yet this is so, for in the chemical treatment the so-called wood sugars, twenty per cent of the total, become soluble. Again, the Masonite process differs from paper making, not only in actually putting lignin to work, but also in saving these wood sugars. They are washed out after the material is exploded from the gun, and Masonite could collect a hundred tons daily. Here is a potential source of twelve million pounds of furfural a year, a lot of pounds of a very useful chemical at a very low cost.

Furfural is a chemical teaser. It jumped into the limelight during World War I, when it was hailed as a valuable waste that might be cheaply recovered from corn cobs. It was also found in oat hulls, so Quaker Oats took it up in deadly earnest as a salvage project. Furfural dissolves a lot of tough customers like waxes, gums, and resins. It is the starting point for preparing many chemicals and it makes some splendid plastics. But it never crashed the commercial gate until a few years ago, when the petroleum industry found it was good for purifying lubricating oil. During the war, furfural in quantity was wanted for synthetic rubber, and a big, new plant was built near Memphis. Cottonseed hulls were commandeered by the War Production Board to stretch the supply of oat hulls. Patents that snarled up the furfural plastics are beginning to expire, and twelve million by-product pounds at Laurel, Mississippi, ought to start something new in Southern industry.

When pine-wood chips are dumped into the big digestors at a kraft paper mill the fumes that rise from this chemical cooking are loaded with turpentine. Two to three gallons can be recovered for every cord of wood, and with turpentine at eighty cents a pound, as it was during the war, this waste recovery became exceedingly attractive. If they did a thoroughgoing job, the kraft mills could recover some twelve million gallons of this sulfate turpentine. That is half a normal crop of gum turpentine.

Mill after mill has been installing the necessary turpentine recovery equipment. Once in, even if the price of turpentine toboggans, this apparatus will not be scrapped. Sulfate turpentine is low-grade, but it is a workable raw material for chemical processing and the naval stores plants have been delighted to have it pumped in by pipe from neighboring paper mills and to buy all they could for tankcar shipments. The higher the price of gum turpentine is held up, the more eagerly will the chemical plants grab this by-product material.

The distinctive waste of the kraft paper mills is "black liquor," a most unprepossessing goo which contains just about all of that waste half of the wood which they take such pains and expense to remove from the cellulose. Most of it is evaporated and burned under the boilers. If, however, it is allowed to stand a few hours, a soapy-oily-frothy curd rises to the top, a fitting cream to be skimmed off this witches' brew of waste. This is tall-oil skimmings or black liquor soap, a mixture of the oils and rosins in the pine wood, saponified by the alkali in the pulp-cooking process.

Tall-oil skimmings are crude stuff, indeed, but they contain a mixture of fatty and rosin acids, valuable in making soap, cleaning compounds, paints, oil and asphalt emulsions, disinfectants, and lubricants. There is plenty of this waste at the mills: one hundred and twenty pounds of skimmings from each ton of wood pulp, which after a chemical clean-up give sixty pounds of crude tall oil.

If all the sulfate mills skimmed their black liquors we could have some two hundred million pounds of crude tall oil a year. Before the war, almost none of the mills bothered, though the thrifty Swedish pulp industry had long found this by-product a pleasant source of profit. In the United States, only a couple of most chemically minded mills, notably West Virginia and Champion, both always keen on by-product recovery, paid any attention at all to this potential market. The trouble was not to recover it, but to sell it. The war changed this. Industrial users who had turned up their noses at this strange, untried material, became eager buyers. Half a score of converting plants were installed—two each in Virginia, Georgia, Alabama, and Louisiana, and one each in Mississippi, South Carolina, and Texas—and a number of other mills began collecting and shipping skimmings to these plants for refining. By the end of the war most of this waste was being salvaged, a total of eighty thousand tons.

Crude tall oil, that is, the natural mixture of fatty and resinous acids as separated out from the skimmings, is not going to find a steady sale in peacetime. Everybody admits that, but there are two thoughts about how much further the refining process should go. Certainly it must go at least through a chemical clean-up stage to remove impurities, lighten the color, and cut down the odor.

Should the fatty rosin acids be separated completely, or how much, or at all?

"Not at all," says one line of reasoning. "Why take out what our best customers will only put back again? In soap, varnish, printing inks, linoleums, and many emulsions, it has long been standard practice to add certain amounts of rosin to the fats and oils used in their manufacture. What we need to do is to produce a fine, light-colored tall oil, containing different standardized percentages of rosin acid so as to meet the exact requirements of the various consuming industries."

The other side counters, "Unless we refine crude tall oil to the point of separating the fatty and rosin acids, we must always sell a cheap product in competition with the lowest-price, off-grade technical oils, plus rosin, which everybody seems to forget has sold as low as two cents a pound. We ought to sell on the basis of our most valuable product, the fatty acids, for those found in tall oil make fine drying oils for paints, and if sufficiently purified, they can be hydrogenated into foodstuffs."

Nobody has yet succeeded in producing this superrefined tall oil commercially, although a lot of people are working on it. Even the big naval stores chemical plants, Hercules and Newport, are interested, for this is along their line and the prize that dangles ahead is tempting. Three-quarters of a million tons of technical oils and fatty acids go into American soap every year;

the coating industries use more; the sidelines are almost innumerable.

Since refining is not as simple as it seems and needs a lot of plant and technically trained brains, it is likely to continue in a few centralized operations. The map of the South is already dotted with these plants, and if the market develops a worth-while price, short hauls only will be required to bring in the skimmings from the neighboring paper mills. Thus another highly potential raw material will be available in the South in tonnage quantities. While these words were being written, a significant news item appeared: "The National Southern Products Company, Tuscaloosa, Alabama, originally designed to produce one carload a day of crude and refined tall oil, is now producing six to eight carloads daily. Plans have been approved for building an additional plant."

10

Vistas in Silk and Ramie

WHEN William J. Hale, in 1934, coined the descriptive word "chemurgy"—meaning agricultural production for industrial uses—the South found itself in the singular position of the *bourgeois gentilhomme* who was perfectly astounded to learn that he had been talking French prose all his life. The South had been practicing chemurgy since colonial times, making its livelihood chiefly out of it. Its two greatest crops, cotton and tobacco, have long been the greatest chemurgic crops grown in America.

Although the birthplace and headquarters of chemurgy in the United States, the South resembled that would-be gentleman rather startlingly. Just as he knew no grammar or syntax, so the South never recognized one of the first principles of growing crops to feed modern factories. Our cotton and tobacco growers have no monopoly on this lack of perception. It is shared by the rubber planters in Malaya, the Argentine linseed growers, the Russian hemp farmers, by all agriculturists the world over.

Embedded in Doctor Hale's concept of chemurgy is the principle behind mass production: the idea that you can make more money by selling a hundred units at a dollar than twenty-five at two dollars each. Put the other way around, ten cents' profit on one hundred sales is greater than twenty cents' profit on twenty-five sales, as great indeed, as a profit of forty cents on twenty-five sales.

That is a very modern notion. Until the use of power machinery made mass production possible, it could not have occurred to anyone. Throughout the long ages the formula for big profits has always been restricted production and fancy prices.

The new principle of amassing a fortune by lowering your price to increase your sales was first tried on an industrial scale only a century and a quarter ago by James Muspratt when he was fighting to sell English soapmakers chemical soda ash to replace natural potashes. Cheap alkali meant not only cheaper soap, but also cheaper paper, glass, and textiles. Seth Thomas and the other Connecticut clockmakers learned this same trick two generations before Henry Ford and his competitors proved it so sensationally by selling more and more automobiles for less and less.

But the old scarcity formula dies hard. It is the basis of all Government plans for relief, price control, and full employment, none of which contains any hint of the truth that lower prices benefit everyone while high prices and high wages benefit only small groups at the expense of all others. The same scarcity formula inspires trade unionism, bureaucracy, and the cartels, today's great triple threat against the true and lasting prosperity based upon an economy of abundance.

Ninety-nine farmers out of a hundred still believe in the scarcity formula: short crops, long prices. That is why the chemurgic idea has had to date such a pitiful record of slow acceptance. By the same token, that is why the gradual emergence of the chemurgic idea as a recognized principle in our farm program will be such a revolutionary force in American agriculture. When a man bites a dog, that is real news; so when the cotton growers of West Texas instinctively talk about man-hours to the acre, not bales to the acre, that is a revolution in agricultural thinking.

Here we have farming in terms of modern industry with the emphasis upon costs, where it belongs in an economy of abundance. Only when the manufacturer and the farmer talk a common language—volume output at low prices, but larger profits—can the producer and processor of chemurgic raw materials understand each other's problems.

Such a mutual understanding would mean much to the South, which is still the great agricultural section of the country, but where the great industrial expansion of the near future will take place. In the South the chemurgic movement is stirring strongly, and every chemurgic meeting held preaches the doctrine of chemical values—that cotton is cellulose; corn is a combination of starch, proteins, and oil—and teaches the new principle of modern industry, the law of abundance. Scores of chemurgic projects are afoot. Some of them are pretty visionary; others as realistic as a microscope. But all of them are significant.

A sales agency at the North Pole for electric refrigerators does not sound a bit more fantastic than a ranch in the wide open spaces of Texas to grow silkworms. The picture of a bronzed, rangey Texan in cowboy boots and a ten-gallon hat herding silkworms is a hilarious fantasy. And yet, faster than a secretary could enter their names, the sane citizens of Mineral Wells, deep in the heart of the cattle country, oversubscribed the stock of a new company that proposes to crack the Japanese monopoly in raw silk, one of the most powerful commercial monopolies left on earth.

This spirited enterprise had its beginning a dozen years ago when Peter Nader came from Syria and settled in the neighboring town of Mingus. His old mother came with him, and together with her few little treasures wrapped in a handmade silk handkerchief, she carried a tiny box of carved olive wood. Within the box were a couple of ounces of grayish brown specks that looked like poppy seeds—silkworm eggs.

From that point the story unfolds as logically as the links of an anchor chain. There were a couple of lucky breaks and three unusual men played star roles in the development of the plot; but it is a straightforward American success story of mechanical ingenuity and commercial courage.

The Syrian colony at Mingus got the first break. They brought with them Near East traditions of silkworm culture and a highly perfected folkcraft in preparing, spinning, and weaving silk fibers. It was a fortunate happenstance that they settled in a region where mulberry trees had flourished for years.

When and where these mulberry trees originally came from, who first planted them and why, are details the local antiquarians debate. Mineral Wells has long been a famous health resort, and the most credible explanation is that the mulberries were originally set out in an outburst of civic beautification some fifty years ago. Anyway, they were there, ready when Grandmother Nader hatched out her cherished silkworm eggs, to provide the best of all foods for the voracious worms that spin the silk cocoons. This was indeed fortunate for the half dozen Syrian families. Neighbors were charmed with their beautiful handwoven scarves and the women were soon busy filling orders that provided a handsome equivalent of egg and pickle money. The fame of their luscious fabrics spread, and the Syrian goodwives began adapting some American ideas to their ancient handicraft.

The silkworms are not turned loose to browse on the

mulberry trees at the mercy of birds and storms. They are parked in shallow trays and one of the first modern conveniences that the Syrian women adopted was the regular American window screen with a two-inch strip tacked around the edges so they can be stacked one above the other. In these shallow trays the silkworms gorge on fresh mulberry leaves for twenty-six days. They need to, for in that brief time they grow to eight thousand times their original size besides storing up material to spin a silk filament from one to two thousand feet long. This filament they weave round themselves in an oval cocoon, some white, some pale yellow, a little over an inch long, from which the sleeping chrysalis emerges as a moth. The female moths lay three to five hundred eggs that hatch out in ten days at a temperature of seventy-eight degrees. Then the cycle starts all over again.

The moths bite their way out of the cocoons and to prevent this cutting of the fibers, the chrysalis is killed. The old way was to steam the cocoons, but another up-to-date convenience is to freeze them in the electric refrigerator, saving a lot of fuss and muss. Once the chrysalis is dead, the cocoons can be kept a long time. Like any other industrial raw material, they can be shipped half way 'round the world, if handled with reasonable care.

To prepare the silk for spinning, the cocoons are soaked in hot water to soften their natural gum; the end of the tightly wound filament is picked up; and the tiny thread rewound upon bobbins. It takes dexterous fingers to perform this long series of delicate operations and not break the filaments.

Just about the time the Syrian colony down in Texas had nicely established this home-grown silk industry, a big silk manufacturer in New York decided to find out why raw silk has never been successfully and profitably produced in the United States. It certainly had not been for lack of trying. His Majesty's colonies of South Carolina and Georgia, Pennsylvania and Connecticut had all offered generous bounties for silkworm culture thirty years before George III in 1768 agreed to pay a handsome bonus on every pound of raw silk produced in his American domains. A century later there was a spirited revival which promoted a wild speculation in mulberry trees, rivaling the tulip craze in Holland, and prices zoomed to fantastic heights. Right on my own lawn the 1938 hurricane destroyed a gnarled relic of that boom, a gigantic, misshapen mulberry, a picturesque landmark and a banquet hall for the robins and catbirds when its big berries were ripe. The core was too rotted to permit its years to be counted, but the family tradition is that it was one of six planted by my great-great-grandfather about the time the kitchen-ell was added to the house in 1820. The story adds that the mulberry boom was pricked in Connecticut—and elsewhere—by the clipper ships, which brought both raw and woven silks from China.

This old, old question had been nagging Walter Roberts for years. He was thoroughly fed up with the prices and policies of the Japanese silk monopolists. So he sent an expert to the Orient. Two years later he had on his desk a regular tome of a report. The gist of hundreds of typed pages was but a summary of the old disappointment: silkworms can be successfully grown in the United States, but silk cannot be profitably produced here in competition with cheap Oriental labor unless the delicate work of rewinding the cocoons upon bobbins can be done by machine. So Walter Roberts went out to find an Eli Whitney, some inventor who could do for silk what the famous tinkering Yankee had done for cotton. He found him in Gustav Beckman, an engineer with the Universal Winding Company.

These two unusual men have joined talents and resources. Beckman, a reserved chap under fifty, has a brain that is a direct, practical tool, like a scythe or a tooth brush, which just cannot function in guesses or promises. His cocoon-winding machine is now more clever than the nimblest fingers since it can pick up and reel on the bobbin several filaments at once, thus eliminating the throwster operations in the silk mills. Convinced by many trials that it works, Roberts agreed to back it; made an exclusive contract with Universal Winding to build the machines; and took over Beckman to run an operating company. Roberts, a big, quick-moving enthusiast over sixty, still had some business problems to solve almost as intricate as the mechanical problems of his engineer.

Where in America do we get silk cocoons, was the first question. It was answered by the third unusual man who has helped this unique chemurgic project, Ernest Mims. A highly successful young businessman who gave his name to an office building in Abilene, Mims was forced by an accident to retire. To keep going he became manager of the Mineral Wells Chamber of Commerce. He brought a practical business touch to his Chamber of Commerce job, which was a break for Roberts, the silkworms, and his fellow citizens.

Roberts, who had heard of the Syrian colony, came down scouting for cocoons, and Mims not only found him a building for his American Silk Corporation, but he organized the Texas Silk Industries to provide the reeling operations with a steady, adequate supply of cocoons. All this might have been done by any smart Chamber of Commerce as a good piece of development work. But Mims followed through. He has become a working expert in sericulture as well as a talking enthusiast on raw silk production. He helped select eighty acres for an experimental mulberry plantation. "Jim-Bob" Mosley rigged up a special tractor plow to open a furrow, set out the young plants, and close the furrows. A volunteer crew of soldiers' wives followed, straightening up the saplings and stomping them firmly in. In terraced land on contours they planted sixty-three thousand baby mulberries in five days.

A woodworking shop was set up to make the window-screen trays. A school of sericulture was organized, and to make good on the offer of C. G. Fairchild, supervisor of Texas vocational rehabilitation, Mims improvised a wonderful dormitory for disabled veterans in one of the church parish houses. Classes of two dozen are sent by the state for six months' courses. A new chemurgic industry has certainly been well rooted, one that opens up a new horizon for the veterans and many small landowners.

A man and his wife can handle six to ten acres of mulberries, which become productive in two years against seven years for pecan trees. The mulberries are kept trimmed in bush form, not higher than six feet, and the leaves are cut from each side of the row alternately, being ready for a second cutting in ten days. The cutting season is seven months, and the silkworm cycle about two months, so a continuous operation can be set up in assembly line style. The worms are clean and odorless and may be raised in any room in the home, in the attic, a converted chicken coop, a barn, or garage. As a smart publicity stunt-hobby, the gardener at the Baker Hotel is raising them in the lobby, feeding them from the mulberry tree leaves grown on the grounds.

Beckman at the reeling plant will freeze the cocoons

the silkworm farmers bring in. A responsible group of eleven silk manufacturers are pestering him for a contract for twenty thousand pounds of raw silk weekly, and he promises to put in additional reeling machines as fast as the supply of cocoons grows.

Profits?—Mims says, "No promises at all. But anyone ought to clear $200 an acre after the third year and it is possible for a real hustler with skill and experience to make a thousand." One thing that tickled my chemical fancy is that already this infant industry plans to salvage its waste product by recovering from the frozen chrysalis an oil eagerly wanted by the cosmetic trade.

Another famous Oriental fiber with more postwar promises than silk is being Americanized in the South. Ramie may not be as romantic and glamorous as the luxurious filaments of the cocoon, but it is the most amazing and provocative of all the natural textile fibers.

To begin with, it comes from a plant that belongs to a fairyland botany. It is a nettle without stings. It grows seven or eight feet tall, not annually, but if cut, three or four times a year. If its roots are frozen it dies, but if the tops only are frost-killed, it makes up for lost time by putting up more shoots and growing faster than ever. This rank grower is a perennial and once properly propagated it need not be replanted for ten years. Ramie is seriously attacked by no insect or fungus. It is the kind of plant a farmer dreams about.

Ramie makes a ton of fiber per acre per year, two

thousand pounds against cotton's two hundred and fifty. The fiber is ninety-nine per cent pure cellulose, so it is the most efficient cellulose factory available in the country. But it also produces a very interesting natural gum and large quantities of pectin, both of which are recovered in cleaning the fiber. The tops contain little or no fiber, but they bear the leaves and yield about three tons of dehydrated cattle feed, containing twenty-five per cent protein. This should be worth $50 a ton, or $150 an acre. It would be hard to find a more glittering chemurgic prospect than this useful vegetable giant.

Ramie fiber is more astonishing than the plant. It is eight times as strong as cotton, three times as strong as sturdy hemp, and most remarkable of all, twice as strong when wet as when dry. The individual fibers are four feet long; when processed some eighteen inches or a foot long. That is infinitesimal compared with the two-thousand-foot silk filament, but flax is about a foot in length, and cotton longer than an inch commands a premium price. Ramie fiber is more quickly and more highly absorbent than cotton or linen, and it dries out faster. It does not shrink. It has a natural sheen like silk. It can be woven or knitted on regular textile machinery either alone or in combination with cotton, silk, wool, linen, and all synthetic fibers. It dyes level and fast to light and washing.

This is almost a vulgar display of textile virtues; ramie's good properties seem too good to be true. Certainly it is fitted for many uses. For its strength, it is used in narrow webbing for parachute harness, and tests prove it superior for machinery belting and carpets. Its absorbent qualities indicate the ideal towel and bandage. Its fineness and luster have made so-called Shantung or China silk famous, and this illustrious material is crude stuff compared with cotton-ramie and rayon-ramie fabrics that are being produced experimentally in several Southern textile mills. During the war, when the water mains of blasted London were put out of commission, flexible pipes up to fourteen inches in diameter, woven of pure ramie fiber, saved the city from fire and pestilence by rendering much better than substitute service. Out of that harrowing experience, rubberless garden hose is a prospect. Among the surprising varieties of postwar applications that are being studied are upholstery, especially automobile seat coverings, marine cordage and fish nets, packing for stuffing boxes, light-weight men's suitings, novelty fabrics for sportswear and evening gowns. And that does not begin to be a full catalogue of proposed uses.

"Why doesn't somebody tell me about these things?" is the immediate reaction of most Americans, and it is quite a shock to learn that ramie was "discovered" by the British textile industry exactly one hundred and forty-five years ago. It has been used in the Orient for centuries. During the past two decades, at least once a year, some chemurgic enthusiast has proposed to do something about commercializing this plant somewhere in the Deep South.

All these brave proposals ended ignominiously, so that in textile and chemical circles noisy shouts of "Ramie! Ramie!" came in time to be ignored as did the peasant boy's frantic shouts of "Wolf! Wolf!" But the wolf did come at last, and now it seems that ramie has actually arrived. War demands and the dislocation of normal raw materials have been responsible for this.

Hemp, sisal, abaca, all the cordage fibers were critical materials. Ramie was obviously a superior substitute. The British found it the best fiber for parachute harness. The United States Navy discovered that it is better than flax for packing propeller tubes. These war demands, however imperative, would not have brought ramie to the market, if this critical need had not prompted the discovery that machines for decorticating hemp and sisal could be adapted to the mechanical separation of the long ramie fibers from the stout stalks. Always this has been the great drawback, for ramie fiber separated by the primitive method of beating the wet stalks was expensive. Moreover, in this harsh treatment it acquired a grievous fault. The sharp blows bruised the fibers, rendering them very tender at this point. Thus ramie had won an undeserved reputation for being a brittle fiber. Mechanical decortication cut costs and eliminated this brittleness.

Until Pearl Harbor exports of ramie from the Philippines grew rapidly. Afterwards Haiti and various Central American countries adopted the plant. Interest in ramie, always slumbering in the Deep South, reawakened. Substantial and influential friends began bestirring themselves, and the states of Alabama and Florida and the United States Department of Agriculture lent a hand.

A prominent banker in Alabama caught the vision of another new Southern horizon and started a ramie plantation. When he ran into processing troubles, the Newport Industries took over at the suggestion of the Governor. It was a logical move, for the company had been operating throughout the Deep South for thirty years, collecting pine stumps for its wood rosin plants at Pensacola and Bay Minette. It has, therefore, local experience and agricultural interest with the resources to back them up and, most important, an experienced, aggressive research staff. In cooperation with the state, experimental plantings were made at the Alabama State Prison Farm at Atmore, and a thorough scientific investigation—chemical and microscopic—was undertaken of the ramie fiber. Simultaneously the mechanical decorticator was studied and an improved machine devised. Samples of fiber, both raw and degummed, were sent to textile mills and spinning, weaving, and knitting experiments started, working with ramie alone and in many combinations with cotton, wool, and burlap. A little later the Newport chemists began exploring the possibilities lurking in the recovered gum.

At the same time Newport made another experimental planting in the Florida Everglades where a lively and promising interest in ramie was stirring. The State Farm at Belle Glade had put in a few acres, and Charles R. Short of Clermont had invented a portable decorticating machine. Later Short set up the first commercial degumming plant at the Florida State Farm. So encouraging were the first field tests in the Everglades that Newport and United States Sugar, owners of a vast acreage near Clewiston, joined forces and have a planting of five hundred acres of ramie which is just coming into production. Also in the Everglades are plantations of the Sea Island Cotton Mills and of Dr. Brown Landone, while the Florida Ramie Products Company, managed by Captain Alexander Kidd and backed by the Johns-Manville Company, has recently purchased five thousand acres of new land. Already there are three practical decorticating machines; one, the Gardiner, especially designed for the small farmer. Half a dozen different processes for degumming, several of which avoid the use of strong alkalies which might tender the fiber, have been perfected.

The Everglades are off to a flying start and appear to have several important natural advantages. Nevertheless all the new plantings are not in southern Florida. Climatic conditions will confine ramie pretty closely to the same belt along the Gulf Coast where its fellow countryman, the tung tree, has been successfully acclimated, and because it grows best in a slightly acid soil, it is not apt to move very far into Texas. The Delta of the Mississippi is tempting territory, and there are several trials going on there and at scattered points in Florida, Alabama, Mississippi, and Louisiana.

After so many cries of "Wolf! Wolf!" one is inclined to caution. Now, however, it really begins to seem that we shall have a workable supply of this most versatile fiber. With silk the South will have two more stout links to its lengthening chain of chemurgic products.

11

Other Chemurgic Projects

THE SCOURGING BLIGHT that has all but exterminated the chestnut has cost us more than we can estimate. It has destroyed one of our most stately, characteristic trees and robbed us of our best native nut. From a strictly dollars-and-cents point of view, because the bark of chestnut was our most important source of tanning extract, especially for heavy leathers, it has created grave problems for the tanning industry and made us more dependent upon imported tanstuffs. The financial blow has fallen most heavily upon the South in a section that can ill afford economic losses.

The mountainous region from the Blue Ridge and the Alleghenies in Virginia and West Virginia, through the Great Smokies to where the mighty Appalachian Range fades out in northern Georgia and Alabama, has been the very citadel of the American tanning extract industry. The chestnut blight has wiped out its most valuable asset and threatens to drive away many tanneries which had been long established in association with local supplies of tannins. These dislocations are most painful in an area where industrial plants are few and far between and where the country, though inimitably beautiful, is much too rugged for profitable agriculture.

The problems raised by our vanishing chestnuts are also national. Within ten or a dozen years, so the experts tell us, there will be no domestic supply of chestnut whatever. It is unwise to be so dependent upon imports of such a vital industrial material as tanning extracts, but it becomes downright improvident when we know that the future supply of the foreign material is itself uncertain. We have long supplemented native chestnut chiefly by quebracho from Argentina and again the experts warn us that in twenty-five years the accessible native trees will be cut down. Since the quebracho is small and slow-growing, the prospects are not good.

Facing these facts, the General Education Board of the Rockefeller Foundation decided in 1940 to make funds available for a thoroughgoing scientific survey of the natural tanning materials of the South. Alfred Russell of the Chemistry Department of the University of North Carolina was selected to execute this good idea. It was a tough assignment, for it meant timber-cruising an area of some 468,000 square miles from the soaking, subtropical jungles of southern Florida to the steep, wind-swept flanks of the highest mountains east of the Mississippi. Along with physical hardihood the tanstuffs hunter must bring to this job a working knowledge of botany; skill and scrupulous care in collecting good, average, representative samples for testing; chemical expertness in the preparation and analysis of tannins; and, since the final proof of any tanning extract is the leather it produces, practical experience in the old, esoteric art of tanning. On all counts Russell made good.

Despite a distressing skirmish with malaria mosquitoes in the swamps of the coastal plain, he completed the survey within two years. He collected two hundred and thirteen specimens; prepared extracts from them; analyzed each and made sample tannins. He has discovered four promising commercial prospects for the Southern tanning extract industry.

Russell is an unusual example of the devout disciple of pure science who could put through a job of this kind with all its rough-and-tumble field work and its definitely commercial objectives, and yet be content with its scientific rewards. I have seldom met a chemist with a more precisely scientific attitude toward his work and less interest in the practical application of his findings.

When I inquired about the future of the most promising of the tannin-bearing trees he had found, he answered lightly, "I wouldn't know. I've got the facts; it's up to the extract people now." And, of course, it is.

"I expected to collect some four hundred specimens," Russell continued, "and no doubt there are many other trees and shrubs in the South that contain more or less tannin, but it is senseless to try to study all these rare specimens. Present supply is inadequate for industrial use, and they are rare because they propagate slowly, so the hope of future wild stocks or of successful cultivation is dim."

Such a practical statement seems to belie his commercial indifference, but then, I found Alfred Russell a most delightfully contradictory man. Maybe that is the result of his background: born in Belfast, educated at Queen's College, Ireland, taught at Glasgow, research at the University of Illinois, more research in Philadelphia, professor at North Carolina; a kaleidoscopic career that would naturally make for diversity. We spent an afternoon together and he showed me his stockroom of samples, a miniature lumberyard in the cellar of the Chemistry Building, and his laboratory, a miniature tannery where two revolving glass drums were sloshing hides in new tan liquors. In his little office with a big table piled high with a collection of test-tanned leathers in the corner, he summarized the results of his prodigious tannin hunt.

"The best chance for a new tannin supply in the South," he said, "appears to be in southern Florida. Here we found two exceptionally likely woods, the Darling plum and the buttonwood. Growing conditions are most favorable, and slow-growing mahogany makes a diameter of twelve inches in ten years. But lumbering conditions are hard and means of transportation are simply nonexistent.

"Neither plum nor buttonwood can be pulped, and as a combination wood for both paper mills and extract factories probably the most promising in the South is the Australian pine. It is widely spread through the tropical and subtropical areas; is a fairly fast grower and seeds quite readily. It would respond to intelligent forestry methods."

For years many appraising, speculative eyes have been cast at two varieties of what in New England we call "weed trees"—the sumac and the scrub oak. My trite questions brought forth some new information.

We import normally some two thousand tons of Sicilian sumac leaves, from which the commercial tanning extract is prepared. Russell and his helpful students made comparative tannins of domestic and imported sumac and found that our so-called dwarf or winged variety is quite as good. The difficulty is the high cost of stripping the leaves from the stems. The Sicilians pick both stems and leaves and by the good old Biblical processes of flailing and winnowing remove most of the stems. This primitive operation has long been mechanized and with scattered plants to do this part of the work, sumac might become an incidental Southern crop. The leaves are gathered at a season when regular farm work is light and a family could collect from eight hundred to twelve hundred pounds of green leaves and stems in an eight-hour day.

As for the scrub oaks, Russell pointed out that they are only regular species stunted by adverse environment. Tapping a little pile of yellow-covered pamphlets, his published reports, he added, "The chemical analyses and tanning tests of all the common Southern oaks are all there."

Again, it is up to the tanning extract people. Later, reading those detailed reports on two hundred and thirteen different specimens all of which contain more or less tannin which makes a more or less satisfactory extract, I thought of the French chemist who years ago made paper in the laboratory out of scores of different forms of cellulose from lawn clippings to oak wood. It can be done, but—

The South's problem of the tannins is universal. Next to the erroneous scarcity philosophy of farmers, the biggest stumbling block in the path of chemurgic projects is the high cost of collecting farm products or by-products in suitable form for industrial use. Paper from cornstalks, for example, is an enticing idea. Cornstalks make good paper and they are available as a waste product in many localities in quantities that add up to vast totals. But in competition with wood, it takes a couple of big truckloads of cornstalks to provide the equivalent quantity of cellulose in a single good log. Cornstalks are expensive to collect and bring to the paper mill. They are bulky and very dirty, increasing the costs of handling and cleaning. Despite their apparent abundance, they cannot be concentrated at one point in sufficient quantity to keep even a little mill running a couple of months, and as it is a seasonal crop, there is no chance whatever of developing an adequate annual supply.

In one way or another, these handicaps almost always appear in every chemurgic program. If they cannot be overcome, disappointment is inevitable. In the case of tannins, coming scarcities will raise prices. This will hasten a practical solution, for it is profits rather than fears that meet shortages of industrial raw materials.

It is no coincidence that our great chemurgic products, such as wallboard from the spent canes of the sugar mills or the chemical furfural extracted from oat hulls, are made where their raw materials accumulate in quantity as the waste products of another industry. Thus the costs of collection are prepaid. At New Orleans a big new plant is one of the neatest examples of this up-to-date chemical art of turning neighboring by-products into main products. Flintcote makes its roofing shingles entirely out of materials that otherwise would be wasted. In this plant over half a million tons of residues from the near-by petroleum refineries and more than two hundred thousand tons of oyster shells are converted to valuable use. That is not pennies in any industrial balance sheet.

In the new Southern economy the mastic roofing shingle is particularly interesting because World War II compelled a switch from one waste to another and encouraged the manufacturers to extract new values out of their own by-products. As evidenced by Flintcote's big new plant, this *n*th degree of salvaging salvage is successful.

Asphalt shingles are built up of a feltlike sheet which is impregnated with asphalt. The top surface is coated with a layer of chips or granules of a nonflammable material. Before 1938 the Flintcote plant in New Orleans made its felt entirely out of rags; not the nice, clean white linen rags that go into the finest writing papers, but the very rag-tag of the secondhand dealers' stocks. This lowest grade of all the discarded odds and ends of the nation's fabrics was a high-grade raw material for them. It made a fine, tough, absorbent base stock. It was cheap, too, but about three-quarters of its cost was the labor of collecting and the handwork of sorting. When the ragman and the rag pickers went off into the Army or to better-paid war jobs, this cheap raw material advanced in price and even at higher prices began to be scarce. Flintcote began working a little wood pulp into their base stock.

In their fine tradition of working up wastes, they did not go out and buy only regular pulping logs, but began to use as much as possible of the branches and trimmings of the neighboring lumbering operations. Their research staff studied the problem, learning fact by fact how to add more and more of this waste wood pulp to their beaters. Today they are using twenty-five per cent rags and seventy-five per cent wood. This is a long way to come in seven years, but they expect to go even further.

Back in the prewar days the petroleum refineries used to pay Flintcote ten cents a load for carting away the final waste residues of their stills. Today the refineries must be paid five cents for this black, tarry, evil-smelling gunk. This somersault in costs is bad enough, but to make it worse this by-product asphalt is not always the good old quality that it used to be.

The petroleum refineries in the New Orleans area are turning out more and more high-test gasoline, and the asphalt residue from the modern cracking process is not the same as the asphalt from the older distilling process. Impregnated in the felt it tends to become brittle and to crack, when laid on a roof and subjected in summer to broiling sun and cooling rains and in winter to alternate freeze and thaw. Like the rags, the older type of distilled asphalt became more expensive and dwindled even to the point where it was insufficient to meet their needs. To a company jealous of the quality of its finished product this was a grave problem. Again the chemists found a solution. By various heat and chemical treatments, they now give the new cracked asphalt the properties required in a shingle.

The only element in the business not upset by the war was the coating granules. These, too, are a waste, the slag from the Birmingham steel mills, ground up coarsely and colored light or dark gray, green, red, blue, or whatever hue architectural fashion dictates. The two hundred thousand tons of oyster shells are finely pulverized and used as a filler in the base stock.

Though founded on wastes that nobody wanted, this business has had to face a series of readjustments in costs and materials. It costs money to modify cracked asphalt and with the radical change from being paid for to paying for this material, the Flintcote executives have had to do some figuring to fill big Government contracts, sell under ceiling prices to the public, pay higher wages and taxes, and earn dividends for their stockholders. They attacked these formidable problems in the modern spirit and with scientific weapons.

The Flintcote manager in New Orleans, William N. Lehmkuhl, is a friendly chap, as alert in business as a cat at a mousehole, but he is a bit of a fanatic, a mild monomaniac on the subject of wastes. It really distresses Bill that the American people get only about fifteen per cent out of all the values in a felled tree.

"We are using second and third growth timber," he told me, "the stuff that is cut out in clearing pine lands, the trimmings that are thrown away in pulp or lumber operations. Constantly we are trying to use more limb stock. By chemical means we are extracting the rosin from our pine chips before pulping them. We have been experimenting with a portable hogging machine that we could drive out in the woods to make our chips in the spot to save transportation costs. During the war we made asphalt emulsions and sent shiploads to Brazil and Africa to stabilize the runways of Army airfields. We are working on synthetic rubber-asphalt tiles and on building plywood made of a felt-base core faced with tulpelo, a nice sort of soft hardwood that works up beautifully.

"You see," he explained with a cheerful grin, "when you've got to pay good money for your raw materials you must get your money's worth, and we are trying to develop some new products to recover that fifteen cent differential in asphalt."

Bill and the new plant he runs both exemplify that new industrial spirit in the South and I was not surprised when he spoke out against low wage scales and did not complain about discriminatory freight rates. He was only echoing the views of many Southern industrial leaders who do not all hold the traditional views that one has been led to believe they cherish religiously.

"We don't want twenty-five-cents-an-hour wages," he volunteered out of a clear sky, "because we do not want twenty-five-cent workmen. Low pay rates mean poor labor. Good pay gets good men, South or North. Southern industry is bound to modernize and that means more complete mechanization, which requires better, more skilled workers. We have them in the South, none better in any section, and they are worth good wages."

I brought up the subject of Southern freight rates and he rubbed his nose and thought a split second before he confessed they are not so bad. "If you have the business and the railroads want it, we find them reasonable enough. I certainly would not call the rates discriminatory. In fairness to the carriers, we should remember that volume of business is an important element in their costs. They can carry a million tons of freight over a thousand miles of track a lot cheaper per pound than they can a thousand tons. From the railroads' point of view Southern industries have had less goods to ship more miles. As we make more goods in the South and our local markets grow with greater Southern prosperity, we will deserve a lower freight rate structure. If we fight for it, we'll get it. You can't get much for nothing in this world, and in business it is a bad deal if both buyer and seller don't profit by it."

This manufacturer sounds like a railroad man, and yet the only active railroader I talked with did not even mention freight rates. He talked sweet potatoes.

Thirty-five years ago Jesse Jackson and the Central of Georgia created a new job in railroading when he became agricultural agent of the first road to employ such a specialist. In 1945 he was named "Man of the Year" in Georgia agriculture by *Progressive Farmer.* In making the first of these honorary awards ever bestowed upon a railroad officer, the editor hailed him a "lifelong pioneer." It is an apt description.

Jesse Jackson was originally hired to bring in settlers. That definite, businesslike task did not wholly satisfy his ambitions. New friends are good, but old friends are best, so from the beginning he went out of his way to help the farmers already living within the territory of his railway. In so doing he lifted himself by his own bootstraps from immigration agent to agricultural counselor and farm ambassador of the entire area. He evolved a new type of constructive service which has been widely copied. His good works for Southern farming have thus multiplied themselves.

It is an added tribute to him that the chemurgic development of the good, old yellow yam as a source of starch was first urged in Mississippi by his colleague, S. A. Robert, agricultural agent of the Gulf, Mobile, and Ohio Railroad. Recently Jackson himself has been campaigning for sweet potatoes. Why he selected this crop and how he has encouraged it, is a model of intelligent farm promotion.

Jackson's idea—sweet potatoes for cattle feed—was especially adapted to farm conditions in Georgia. For three years his road has backed him in this campaign by offering county prizes of one hundred dollars for the biggest yields of this crop per acre. These prizes were paid for tons of potatoes grown by an easier and cheaper way of planting, to demonstrate that small pieces of "strings" or ends of potatoes can be planted by all farmers, and so make of the sweet potato a field crop, planting early and harvesting late to make as much weight as possible.

In Georgia, and neighboring South Carolina and Alabama, too, where it is impossible to make a big corn crop, sweet potatoes are the best chance to grow a carbohydrate feed to support the rapidly growing beef and milk businesses. Again the conflict between scarcity ideas and chemurgic yields had to be battled through.

Sweet potatoes for the table trade means the nice, round, smooth-skinned type that sells for eighty cents a bushel up. At such prices a yield of seventy-five bushels is well and good. But for feed or the industrial extraction of starch, sweet potatoes must sell for twenty cents a bushel. Yield per acre must be stepped up into tons, six to nine tons, two or three hundred bushels or more. Good looks do not count when a sweet potato is to be shoved into a shredder and put through a drying machine. In fact, to get ton yields exactly what is needed is the big, ugly jumbos, crooked and nubbly, the very kind the grocery trade disdainfully rejects.

When you are talking sweet potatoes all the time, it takes a bit of showmanship to interest the average farmer in twenty cents a bushel against eighty. Those substantial money prizes for one-acre plots did that kind of talking. Jackson tempted Georgia farmers into growing jumbos. They learned that it is less work.

They proved to their own satisfaction that three hundred bushels at twenty cents is a lot more profitable than seventy-five bushels at eighty cents, though in each case the gross return is $60. Jackson had shown the sweet potato farmer how to cut costs so that bigger yields paid better.

The sweet potato we eat is not a tuber like a white potato, but a thickened section of root. To get fine, round specimens, it is necessary to grow plants in a hot bed, and set the little slips out in the field. All this handwork piles up costs and labor troubles. As long as cotton was profitable, nobody was much interested in sweet potatoes. But with cotton fading out and cattle coming in, the jumbos, which make twice as much carbohydrates per acre as corn, become more and more attractive. Moreover little chunks of root can be drilled by machine just as chunks of Irish seed potatoes are planted. The low-cost-big-yield kinks have not all been smoothly ironed out, but Jackson is enthusiastic over the prospects of a substitute for cotton on the coastal plain and he is too shrewd a veteran to follow a cold trail for more than three years.

To the weight-and-waistline watchers, starch is a particularly insidious type of food. All of us think of it as breadstuffs and puddings. We forget the starched clothes and are surprised to learn that to manufacture pastes and sizing materials and for textile finishing and the commercial laundries, we make nearly a billion pounds a year and then have to eke out with another half billion pounds of imported cassava, sago, and potato starches.

Figures like that rouse any chemurgic enthusiast. Most of our domestic starch is grain starch from corn. Nine or ten million pounds is root starch regularly provided from cull white potatoes. Like the many vegetable oils, the various starches are almost interchangeable, but certain ones are first choice for particular purposes. In the textile industry sweet potato starch commands a one-cent premium over corn, and one of the earliest chemurgic ideas was to grow enough sweet potatoes to declare our starch independence. It seemed a feasible notion. Three hundred or more bushels of jumbos contain more starch than any other temperate zone plant can produce on an acre. The project stumbled repeatedly over the two familiar obstacles: the costs of seasonal operation and the difficulty of securing enough sweet potatoes to keep even a modest plant running at capacity during as short a period as three months.

The real test came during the depression. Relief agencies were hunting any likely agricultural-industrial project, and a federal grant was obtained for a sweet potato starch factory to be located in an area of cutover timberland. Laurel, Mississippi, was selected and a commercial-scale plant installed. The process worked and was improved considerably. A good starch was made and sold at a premium price. The by-product pulp found a ready market as cattle feed. However, the supply of raw material fell far short of providing a profitable volume of business. The Laurel location was not the best, for the yield of sweet potatoes in the area ranged all the way from zero to three hundred bushels. Since sweet potatoes keep badly, there was no opportunity to store sufficient stocks to provide for anything like a year-round operation.

Over at St. Francisville, Louisiana, Douglas Warriner ran into the same stone wall. Here it was certainly not the fault of the location. Warriner, the son of one of the pioneer sweet potato growers, was in the heart of the great sweet potato country centering about Opelousas and Sunset. In Mississippi and Louisiana dehydrating promises salvation. During the war both the Laurel and St. Francisville plants began dehydrating for Army rations and postwar they have a multiple choice: dehydrated for food, cattle feed, or starch. In these sections the crop will doubtless be picked over, the table grade going to that fancy-priced market and the culls to the dehydrating-processing plants.

Out in Lubbock County, Texas, thanks to the example of an agricultural trail blazer, C. D. Elliston, and his record yield of seven hundred bushels, another sweet potato country is opening up. His secret was good seed. Having observed that cotton and grain sorghum planters paid fancy prices for seed, he paid $66 for three bushels of Red Goldens, and his results have jumped the acreage in his county from next to nothing to over five hundred acres.

Down in the Florida Everglades is the newest, biggest project of all: twelve thousand acres and a starch plant to handle seven hundred and fifty tons of potatoes a day. It is backed by the United States Sugar Corporation, it is said, to the limit of six millions. Dr. F. H. Thurber, who was in early charge at Laurel, is helping out technically and here the positive advantages of location are counted upon to provide yields for a large factory. The Everglades soil is rich and moist. A growing season of three hundred days will be utilized to the utmost by staggered plantings. This company goes all out in a new project only after scrupulous scientific investigation. It follows through with thoroughgoing research. With ramie and lemongrass which are also being grown here, Clewiston is likely to be written in big capital letters across the South's new chemurgic map.

In the meantime, a brand-new chemurgic use for sweet potatoes has been found by a Russian-born chemist, Dr. Paul Kolachov, who directs the research for the Seagram distilleries in Louisville, Kentucky. That one is not hard to guess, and this exceedingly resourceful scientist reports that sweet potatoes and grain sorghum are both more efficient producers of alcohol than corn. The possibilities of fresh intersectional agricultural upsets in new uses of old crops is a fascinating speculation, but the new competition seldom comes so quickly that there is not plenty of time for readjustment.

If the apex of chemurgic accomplishment is to create wealth from waste, then the prize with palms and oak leaves goes to a lean, gnomelike, russet-skinned man with a long nose and bright blue eyes under heavy brows who works busily in a large, rather messy laboratory at Texas Tech. Ten years ago Charles G. Rook set himself the task of salvaging the burrs of the cotton plant. These tough, horny bolls within which fiber and seed come to maturity are the ultimate waste—the squeal of the pig—of the cotton industry. The gins have been delighted to give away burrs to any farmer who would haul them off to spread on his fields and return some organic matter and a bit of potash to the soil. A few very provident growers do this, but most farmers think the fertilizer value does not pay for the time and labor, so the burrs generally get burned under the boilers. A fair-sized gin will collect some two million pounds during a season, which make a rather imposing trash heap.

Rook worried about these heaps and the scanty plant-food and low fuel values of the cotton burr. In a speck of a town, Gladewater in East Texas, he had a snug little business of his own making waxes, insecticide bases, naphtha soaps, and other chemical specialties out of crude petroleum. In his miniature laboratory he began analyzing burrs on his own time.

He burned the burrs and found that the ash contained forty-three per cent potash and over three per cent of phosphorus—a respectable content of plantfood elements. He learned that the raw burrs contained a lot of glucosides and pentosans, from which furfural might be prepared. He extracted a workable tannin. He discovered that after these chemical portions had been removed the remaining fiber could be pressed into board without the addition of any binder or that by treatment with a single cheap chemical this fibrous material became a good plastic. In the end he piled up such a store of definite knowledge about this lowest caste of all the cotton crop that Texas Tech and the state clubbed together a joint grant to set him up in the college laboratories with an opportunity to test out his economic analyses on pilot-plant scale.

Rook has worked out two possible commercial combinations. From the first he gets furfural, tannin, and wallboard; from the second, tannic acid and a potash fertilizer. For either he figures that one could pay $4 a ton for burrs. At this price the ginners reckon that at least he would never have to shut down for lack of raw material.

Seeking chemurgic opportunities in a test tube fascinates exploring scientists, but it is fussy, painstaking work. Hunting them among the wild plants in the wide-open spaces makes quite a different appeal. This, too, is a promising branch of chemurgy.

Why should we remain content to go on cultivating the food, fiber, and other crops handed down to us by our forefathers or picked up from native races? Why should we not go out and explore the vast unknown wealth of botanical resources? Why not, indeed—there are over twelve thousand distinct plant species in the South!

In this far-flung field the favorable climate of the South multiplies the chances of success and the semi-desert regions of the Southeast and the semitropical sections of Florida and the lower Gulf Coast are especially inviting hunting grounds.

Two leaders among the plant hunters are linked by the unusual coincidence that their first names are the same—not John or William, but Cyrus—Cyrus L. Lundell and Cyrus N. Ray. After that, save for their zeal in ferreting out new plants that we might put to work, the resemblance ceases abruptly. Lundell is a slender, rather quiet, self-contained scientist; a trained, exceptionally able botanist; a scholar; for all his enthusiasms a circumspect and cautious speaker. Ray is a stoutish, baldish, hail-fellow-well-met chap, a physician who rides his hobbyhorses of Southeastern anthropology and natural history hard and long. He is an amateur, an exceedingly well-informed amateur, but overflowing with the vigor and go of an unbroken colt. Ray will talk your ear off as he drags you from one strange cactus to another in the little garden-museum behind his home in Abilene. You must coax out of Lundell by patient questioning half of what he is doing as director of the Institute of Technology and Plant Industry at the Southern Methodist University.

Ray is convinced that we are passing up two great fiber plants in the agave, which the Mexicans use in making their potent drink pulque, and in the nolina, a rank-growing, grasslike plant of many species. His darling is the sotol, a vegetable jack-of-all-trades. It belongs to the lily family and its starchy bulb is edible. It has a treelike growth with a trunk that in some species is three feet in diameter, which is a source of alcohol. Its leaves, sometimes six feet long and generally about half an inch wide, are used by the Indians to thatch roofs and make baskets. Maybe Ray has something there.

At the time of our rubber crisis the Government sent Lundell down into Mexico exploring for latex-bearing species. He found some—none exceptional—but he also brought back a number of credible commercial prospects. These he is cultivating in an experimental planting of chemurgic novelties. Some of his most promising candidates are the garcia, an evergreen tree from Mexico, easy to grow and producing an oil said to surpass tung in its drying properties; the jojoba, whose edible seeds produce fifty per cent of an oily wax used during the war in shoe polishes, and which is already being cultivated on a six-hundred acre plantation in Arizona by Durkee Famous Foods; the chilte tree, kin of the bull nettle, touted as rubber-bearer, used in making chewing gum, with promises of industrial worth for a hard resin and a drying oil that it produces.

Despite his calm exterior, Cyrus Lundell burns with a consuming enthusiasm for the future of highly specialized crops, specialized as to their employment and to the soil and climatic conditions of various sections of the country. He is carrying on breeding experiments to produce not a bigger, better corn, but a more prolific corn adapted specifically to the blacklands of Texas in specialized strains to satisfy the distinct requirements of the meal miller and the alcohol distiller. His faith declares itself: "We face a frontier different from and more promising than any our pioneer ancestors knew. We need research activated by the pioneer spirit of our forefathers, and business pioneering—capital that visualizes the opportunities and possibilities of the South."

12

Mineral Ores and Water Power

THE SOUTH'S spacious storehouse of raw materials is so surprisingly stocked with metals and minerals, it dumbfounds even the boasters of the old chamber-of-commerce school. We do not commonly think of Texas as second among the states of the Union in the output of minerals, and few would guess that West Virginia stands fourth, or Kentucky, ninth.

Each year nearly a third of the country's total mineral wealth comes from the thirteen Southern states, adding some two and one-half billion dollars annually to the income side of the South's budget. Southern production of several minerals conspicuously important in modern industry dominates the market. All industrially important minerals produced in the United States are found in greater or less quantities in the South.

Sulfur, the source of sulfuric acid, that old war-horse of all chemical operations, and helium, the lightweight, nonflammable gas so essential for lighter-than-air aviation, are strictly Southern monopolies. Phosphates, one of the three essential plantfood elements, come almost exclusively from Florida and Tennessee. For the metal aluminum and the chemical alum, we depend upon bauxite from Arkansas, Alabama, Georgia, and Virginia. The newer white pigments for paint are based upon titanium and barium found in Virginia, Tennessee, and Florida. North Carolina produces more mica—essential in electrical insulation—than all other states combined. Texas leads in fuller's earth and Georgia in ochre and other high-grade clays.

The Birmingham district has long been famous for its unique combination of iron, coal, and limestone, the steel-making trio. From great, modern Texan refineries come two imported metals: tin from Texas City and antimony from Laredo. And most significant of all, the country's producing centers of aluminum and magnesium, the lightweight metals of tomorrow, are both in the South.

For ages weight has been associated with worth. A ton of bricks, sugar by the pound, the two-carat diamond: the greater the weight, the higher the value. Almost instinctively we have come to attribute to the force of gravity desirable characteristics of durability, substantiality, permanent value. The great bulk of the Pyramids, the massive granite arches of the Roman aqueducts are impressive symbols of this idea. The ponderous, carved oak sideboard loaded with heavy silver is the very hallmark of pre-eminent respectability. The clodhopper hunting boot and the blanket-like ulster express the same thought. But quite recently we have acquired a new standard of values.

"Some day we shall free ourselves from the load of gravity by appreciating the merits of levity." That prophecy was made twenty-five years ago by the man who led the fight to make magnesium so plentiful and so cheap that it might become a common structural metal.

It was in May, 1919, the spring after the First World War ended, and Herbert H. Dow and I were sitting, after lunch, in comfortable rustic chairs in his Michigan garden. In the distance stretched his experimental orchard, row after row of apple trees in full bloom, ordered clumps of pink and white blossoms beneath billowy white clouds against a sapphire sky. The peaceful beauty of the spring afternoon enveloped us and that big, benign engineer, so vigorously exacting as a manufacturer, so militant as a competitor, relaxed and became an industrial philosopher.

"It takes twice as much energy," he said, "to move two pounds as it does to move one. Slowly we are beginning to learn that by the right choice of materials and through better design, lightness can be achieved without sacrifice of strength or durability. Why waste both energy and materials struggling against gravity? After all, levity has positive virtues. It is certainly a great convenience.

"Once we get this idea we will be off on a new line

of progress. The changes will be as great as those that followed the first use of metals. It is a really new idea, much more revolutionary than communism, or disarmament, or relativity. It is a fundamentally different conception of value, and I don't dare dream of what it will mean to transportation, to our ways of living, to our whole economic system; what its effects will be upon our political, legal, and social structures."

This change in our thinking is rapidly coming to pass. The airplane has hurried it. Already we have streamlined many old conceptions. Clothing, tools, automobiles have all acquired the positive virtue of levity that Dr. Dow foresaw. The prime specifications for a radically new type of lawnmower planned for tomorrow are that it shall "cost not more than ten dollars and weigh less than ten pounds." Postwar models of oil-well derricks and harvesting machines, typewriters and vacuum cleaners, picnic hampers and eyeglass cases, and of scores of other products, all feature this new attraction of light weight. This recently approved virtue is at once the most insistent reason for and a most pleasing result of our constantly increasing use of plastics, plywoods, and the featherweight metals, aluminum and magnesium.

The South is ready with all these buoyant materials for fabricating into goods to conform to our weight-saving ideas. One of the astounding miracles of the war was that what threatened in 1941 to be a critical shortage of light metals needed for fifty thousand warplanes became a surplus production demanding cut-backs in 1944. The South has the capacity to produce a million tons of aluminum and three hundred thousand tons of magnesium; a greater output of both than existed in the whole country prewar.

The biggest magnesium plant in the world—that of the Dow Chemical Company at the mouth of the Brazos River in Texas—taps the infinitesimal but infinite supply of magnesium in the ocean. Headquarters of aluminum production in the United States—Alcoa in Tennessee and Reynolds in Alabama—are at the geographical center of the country's bauxite supply. Threatened by voracious war demands, that supply has been multiplied many fold by a new lime-soda treatment that extracts high-grade raw material out of low-grade ores formerly considered useless. At Listerhill and Hurricane Creek, Alabama, at Baton Rouge and Mobile, are four big, new, war-built plants to produce alumina: not aluminum, but the processed raw material from which both the metal and the alum are made. Another alumina plant at Harleysville, South Carolina, is a wartime experiment in coaxing this valuable aluminum oxide out of clay, a source that rivals the ocean. There is little worry now that we shall run short of either of these metals.

Quite the reverse worry arises—where are we going to put all the lightweight metal we have on hand? Immediately the scene shifts to Washington, for the Government built and owns the excess-capacity plants and is under no obligation to sell them to their wartime operators. In fact, an effort is being made to encourage competition. In each metal a single company, Alcoa and Dow, now occupy a dominating position, won during the lean years of the depression when what now seems their pitiful output was much greater than the country wanted.

Though the present outlook is so cloudy, the very men who ought to be lying awake nights stewing show few signs of insomnia. If they do have trouble getting to sleep, I suspect it is because their heads are buzzing with ideas for the future. When I asked the supposedly embarrassing question of my old friend, "Dutch" Beutel, general manager of the Dow operations in Texas, he looked out of the window at the big plant the Government had erected alongside the original Dow unit and shrugged his shoulders.

"Come now, I'm only a production man," he said, showing his strong white teeth in a silent laugh, "and that is strictly a problem for the management and the Government, chiefly for Uncle Sam, I reckon. I want none of it."

Whether that "it" referred to the problem or the plant I could only guess for he tossed across his desk a three-inch cube of metal and asked if I had seen this new alloy of zinc and magnesium. And he was off, extolling other new magnesium alloys with copper, tin, and aluminum, as well as zinc, and reeling off their merits like a country auctioneer warming up over a tractor and all its appliances.

"And don't you ever forget," Dr. Beutel concluded, "that at twenty cents a pound magnesium, the lightest of all metals, is cheaper on a volume basis than either copper or aluminum."

At Listerhill, R. S. Reynolds luckily swooped in on his private plane between his winter home in Miami and the Reynolds Metals Company's office in Louisville. He was once dubbed "Aladdin of Aluminum"—alliterative, but inept, for it smacks of magic. Though he has pulled some pretty rabbits out of the hat, he is as much of a sleight-of-hand artist as the village blacksmith.

During the evening we spent at the Company's snug guest house on the shores of Wilson Lake with Basil Horsfield, vice-president in charge of aluminum production, and half a dozen others of the staff, Reynolds impressed me time and again as an American industrial edition of the idolized Little Corporal of France. Physically he is like Napoleon, a big-little man, stocky, baldish, teeming with vigor and confidence. He has other Napoleonic traits. He has fought many a campaign all over the map of industry, and his victories have been won by the swift-moving tactics of surprise attack. His associates, like the Old Guard of France, will follow their spirited leader into the very jaws of Hell. As every

French soldier carried in his knapsack a marshal's baton, so every Reynolds employee knows that executive titles—and salaries—can always be won by promotion from the ranks.

Richard Samuel Reynolds got into aluminum naturally. Born into the tobacco business at Bristol, Tennessee, he capitalized his connections thirty-five years ago by starting to make lead foil for packaging pipe tobacco. Aluminum foil for cigarettes was a natural growth, but he went on and sold metal foil for protecting chewing gum and candy. In the old days he bought his aluminum stock, but shortly before the war he literally began "rolling his own." Today the Reynolds Metals Company makes more aluminum than the whole of England and France.

Where's all this aluminum going? "R.S." has no qualms on that score, and that evening he had his own associates, scientifically minded chemists and realistic plant operators, on the edges of their chairs as he exploded idea after idea. Aluminum foil with the labels all printed on it, as cheap as waxed paper. The Army has shipped everything from ammunition to Bibles in aluminum and found it good. Bread wrapped in hermetically sealed aluminum sheets keeps oven-fresh for weeks. Aluminum foil inside the roofs and walls of our homes is a perfect insulating material. Steel plated with aluminum becomes everlastingly rustproof. A pound of aluminum makes 11,300 yards of metallic thread that can be woven with cotton into amazingly soft and beautiful textiles. If you laminate aluminum sheets with plastics and plywoods, you have a whole new series of materials. What about aluminum shingles, lightweight, fireproof, and insulating all at once? Aluminum freight cars would save half the nation's freight bill. Aluminum kitchenware is familiar, but what about aluminum tableware, furniture, fountain pens, costume jewelry, or what have you? We tumbled into bed that memorable evening at one a.m. to dream of the coming Thistledown Age of American Civilization.

Reynolds expects to keep all his war workers who want to stay on the job, re-employ six thousand returned veterans, and then hire some more. He is confident that at the present price of fourteen cents a pound, there will not be enough aluminum to go 'round and that every penny shaved off this price by better techniques and bigger output will send the demand zooming.

"We'll make more and more jobs," he declared emphatically, "by producing more and more aluminum. What we are going to need is more and more electrical power."

That optimistic statement raises big questions. Power is an asset as tangible as gold, and hydro-electric power is one of the South's great natural resources. Hence the Tennessee Valley Authority, the largest producer of electrical power in the South, has enormous potentialities—for or against—Southern industry. Specifically, the

TVA may well determine the future of three of the South's greatest industries, the old, well established hydro-electric power and fertilizers, and the very new, very potential aluminum.

These industries are rooted in three most important Southern resources: water power, phosphate rock, bauxite ore. Their products—energy, plantfood, metal—are so basic that they have a tremendous influence upon the future development of the entire South. They involve the lives of millions of Americans who live far beyond the confines of "The Valley."

TVA's well advertised successes—which are sharply questioned and which have not been unequivocally proved—are stimulating proposals for similar River Valley Authorities in the South and elsewhere in the nation. Therefore its accomplishments, from the point of view of their industrial potentials, should be regarded neither myoptically nor through rose-colored glasses, not only for the sake of Southern industrial development, but also because they affect the economic future of the whole nation.

Amid many splendid accomplishments two failures of the TVA rise as bold and uncompromising as a pair of gigantic smoke stacks towering high above a beautiful modernistic group of factory buildings. The first is the failure to meet one primary objective for which the Congress set up this magnificent experiment in regional planning. The second, which grows out of the first, is a failure to employ the primary asset of the TVA most effectively towards the ends sought.

The TVA was to provide a yardstick to measure the rates charged for electrical power by the privately owned public utility companies. Interpreted literally, as we, the people, supposed it to be, this yardstick was expected to check not only the comparative operating efficiencies of public and private enterprise, but also their respective public service in the social meaning of those words. This yardstick idea and the promise that this ambitious project would be self-liquidating were the most pressed arguments for the establishment of this enormous publicly owned public utility. They made a strong appeal to many who instinctively disliked or mistrusted this major invasion of the field of private enterprise by the Government. It certainly influenced many of the votes in Congress, possibly sufficient votes to have secured passage of the bill.

That this yardstick has proved to be useless surprises nobody who had considered this proposal at all realistically. But it is most disappointing to many who are deeply interested in the economic good health of the South that the TVA bookkeeping is so obscure and its activities have become so diversified that it has made no constructive contribution to power economics and has so far been little help to forward industrial planning.

The expectation of more practical results may have been incorrigible wishful thinking. However, it was sincere, and there is no doubt that, if the will to make a fundamental, factual contribution to some of our most perplexing problems had inspired the TVA experiment, a priceless yardstick of economic and social values might have been created. Public utility bookkeeping is terrifically complicated, but for years municipal, state, and federal power commissions, private and public accountants, have all wrestled with its technicalities. Standards have been set up and rules laid down. It is perfectly possible to find the financial facts honestly and to present them simply. The TVA could have done so, can still do so. If the Congress really wants a yardstick, we can have it. For comparison a veritable mountain of public service corporation reports is ready to be measured.

During every minute of its existence every private enterprise—Joe the bootblack or the General Motors Corporation—must do two things that no enterprise of the Government, be it the dairy inspector of Cross Corners or the United States Army, is ever called upon to do. It must meet competition and it must count the cost. The voter exercises a very remote control over the management of public business compared with the imperative, immediate orders of the customer to the management of private business. The TVA certainly is not responsible for these conditions, but because it has to operate under them, it scored its second failure.

The TVA was created by Congress to benefit the people of the Tennessee Valley by means of objectives definitely stated: navigation, flood control, reforestation, fertilizer manufacture, and power generation. All of these worthy objectives have been attained, some of them with notable success; the price paid has been enormous, a total well over $900,000,000. Though a billion dollars might be considered fairly liberal, the great failure of the TVA is not its dollars cost but that these vast sums have not been invested intelligently to achieve the ends sought and to serve the needs of tomorrow.

Flood control was certainly one of the primary objectives. The TVA reservoirs have drowned out some seven hundred and thirty thousand acres of fertile bottomland, an area roughly equal to Rhode Island. More acres are permanently under water than the Army Engineers estimate would ever be covered by a flood so great that it might be reasonably expected only once in five hundred years. Total losses in crops to the farmers and in local taxes to the towns of the Valley have never been cast up, but in 1941 the Tennessee Farm Bureau estimated that the annual value of food crops alone formerly produced on the 561,000 acres then flooded by TVA was $13,415,300. Balance that against $1,784,061—official estimate of the Army Engineers of the total average yearly flood loss, and the plain horse sense of the average American suffers a terrific shock.

Elaborate schemes to achieve simple ends, the stable

burned down to roast a pig; enormous effort to little effect, the mountain that labors and brings forth a mouse; tremendous expense for pitiful returns, the ten-ton truck hauling a bag of peanuts—these are old, familiar errors of human judgment. We cannot stop making such mistakes, but we can check and correct them by counting the cost.

While the simplest common denominator is money, we must admit that cash values do not always give the final answer. Nevertheless, even if we toss aside the initial cost and current maintenance expense of the TVA flood control program, we are forced to conclude that what was planned to benefit the people of the Tennessee Valley has actually hurt them. We could afford to laugh at the ridiculous result of drowning forever more land than would be flooded by Nature once in half a century, provided Congress stopped such expensive jokes at the taxpayer's expense. It is no laughing matter, however, that a national experiment in the coordinated economic development of a large region has gone so far askew in one of its simplest, most direct objectives. To the people of the Valley, whose basic problem is low income, this is a tragedy. To the people of the United States, seeking help in the solution of desperate economic problems, it is pretty dismal comedy.

If the TVA has failed to provide a yardstick for hydro-electric power costs, it gives the Congress and the people an exceedingly useful measure for counting the costs—all the costs—of similar experiments. Does the water-borne commerce of the Tennessee River justify the investment and expense of the TVA program to maintain it as a navigable stream? What have been the costs, the real costs, in realistic results as well as taxpayers' money, of reforestation, housing projects, TVA publicity, phosphate mining, fertilizer manufacture, or any other of its many and growing activities?

Hydro-electric power is the heart of the whole TVA project. Its two million kilowatts are its greatest, most costly asset. Its major objective is to employ this power for the best benefit of all the people of the Tennessee Valley. This big block of power is an unparalleled opportunity.

Even jaundiced opponents of the TVA admit that power rates to the individual consumer on farms, in hamlets, in big and little cities have been lowered. What they protest so vehemently is that this saving is but a trifle of the incomes of the people served, an utterly insignificant personal gain that cannot be justified by the mountainous power investment, and that these lower rates to individuals have been an expensive personal gift from their fellow taxpayers.

Let us shelve all these dollars-and-cents arguments both of critics and defenders and count the costs of TVA power in quite a different way.

For years the Tennessee Valley was an acute example

of the South's basic economic problem, the low income of its inhabitants. Being predominantly an agricultural region, when the TVA deliberately destroyed six-sevenths of the fertile bottomlands it cut deeply into what was already a submarginal cash intake.

Everything possible is being done to help the Tennessee Valley farmers solve the problems of hillside farming on washed-out soil. Phosphate fertilizers and lime are being distributed generously. Forestry projects are being vigorously pushed. Still the approach to the problem seems to be out of focus.

With much of the best land obliterated, the prospect of raising the productivity of the remaining land to a point where it will actually raise the income of the community is not good. Even if such an agricultural miracle might be consummated, common sense indicates that the costs measured by economic results will be fantastic.

Rather than attempt to prop up the thin and narrow agricultural base, the people of the Tennessee Valley must be helped to plant their feet firmly upon a broader, more solid economic foundation. The alternatives are plainly migration or industrialization. To the latter end the enormous power output of the TVA operations offers a ready-to-hand, practical means.

Failure to adopt this straightforward solution is not wholly chargeable to the Authority. When the Congress set up the TVA in 1933, it specifically instructed that power should be sold to farmers and to municipalities at the lowest possible rate. Power for industry was to have secondary preference. This restriction then appeared to be reasonable but, from subsequent developments, it appears that Congress should cancel this proviso in order that the energy created by the Tennessee River may be used primarily to furnish jobs for the people of the Tennessee Valley. If TVA power might be sold to industries in quantities they need and at prices they can pay, the opportunity would be created for these people to employ many skills at wages that would immediately raise the per-capita income.

Throughout the Valley it is a threadbare joke that Muscle Shoals is the only place on earth where you can shoot quail from paved sidewalks. These relics of a city that never existed save in the fertile imaginations of real estate promoters date back a dozen years before the TVA was created. Those concrete strips across the open fields are no criticism of the Authority, but they should be a warning. They are monuments to an industrial mistake.

Years before 1918, when the Government built the first power plant at Muscle Shoals, the same mistake had been made by one of the country's great hydro-electrical pioneers, Jacob F. Schoellkopf, who was a courageous, far-sighted, rugged individualist, the very antithesis of the starry-eyed visionary. Yet when he harnessed the power of Niagara Falls, he dreamed of a great industrial city with scores of textile mills, shoe factories, woodworking plants. Like the aimless sidewalks of Muscle Shoals, row after row of great maples that he planted across the open fields behind the gigantic plants that now string along the Niagara River bear silent testimony to his city-planning that came to naught.

What Schoellkof and his business associates did not foresee was that cheap power is not a magnet to attract light industry. At least it is not sufficiently strong to offset a favorable labor supply, proximity to raw materials or to markets, or any of the other logical reasons that commonly determine the location of a plant. But big blocks of cheap power do attract those industries that actually consume electrical energy in their processes, and within a couple of years cheap power made Niagara Falls the American electrochemical and electrometallurgical headquarters. It remains to this day the country's great center for carbide and ferroalloys, abrasives and refractories, alkalies and chlorine.

This logical element in industrial practice has put the Authority in the uncomfortable, and unwarranted, position of being damned if they do and damned if they don't. Inevitably the best progress made in attracting industries to the region has been among the same types of operation that flocked to Niagara Falls. They are consumers of big blocks of power to process the available mineral resources, in this case phosphate rock and bauxite ore. The Monsanto and Victor phosphorus operations and the Alcoa and Reynolds aluminum plants are all big. Some of TVA's most cordial friends—to whom size is presumptive proof of any corporation's anti-social guilt—have criticized the sale of chunks of power to a few large customers.

If jobs by and through power is the best way to serve the Valley, friends and critics and the TVA, itself, must recognize that the first step is to lay down a backlog of big kilowatt consumption by heavy industry. Such enterprises must be big and to survive they must buy electrical energy at a low rate. Once firmly established, their products will draw in fabricating industries which use power to turn many motors but which can afford to pay a higher rate because their products consume relatively little electrical power per pound. In this way, within a very few years, the Tennessee Valley might become a humming center of industrial activity, employing important Southern resources, providing its people raised standards of living, supplying the entire South with industrial materials needed for many kinds of consumer goods.

These glittering generalities can be pinned down by figuring the jobs and the products to be gained through industrial use of say, six hundred thousand or eight hundred thousand kilowatts of the two million available in the TVA system. Used primarily in such heavy industries as would naturally be attracted, this amount of power would sustain an annual payroll of at least fifty million dollars. In addition, such an industrial growth in the Valley would initiate a trend of events that would double or treble this basic payroll in services and supplementary industries.

Aluminum and phosphorous products are already established in the region. With limestone, coal, and labor plus low-cost power all available, it is not visionary to foresee great industries based upon carbide all the way from Chattanooga to Paducah or preferably to below Florence. From carbide, acetylene gas for welding could be distributed by water transport. The Tombigbee Canal connects the Tennessee River practically at Florence with the port of Mobile, whence imported chrome and manganese could be brought in for the production of ferroalloys. These steel-making ingredients could still reach Pittsburgh and Chicago by water transport. Adjacent to Sheffield is low-grade, high-silica bauxite, ideal raw material for the manufacture of synthetic abrasives of the general type of silicon carbide. The by-product alloys are wanted today at the Birmingham steel mills. Salt, sulfur, and petroleum, all gathered in by barge, are the requirements for employing TVA kilowatts in an electrochemical development that might threaten the supremacy of Niagara Falls.

These are not dreams, but they cannot come true unless the power policies of the TVA are revised to an industrial base. Judged by the common-sense yardstick, the benefits to be won by more jobs and new products through combining cheap power with the natural resources of the region appear to be the direct way of achieving practical results from the power, the flood control, the navigation comparable with the staggering total of our TVA investment.

Hydro-electric power in close association with mineral ores is one of the South's most potential sources of wealth. Whether this rare, valuable combination is to be capitalized by the Government or by private industry is a question fraught with vital consequences to the people of the South. The TVA has not given us a clear-cut answer. In fact, it has hidden the costs and befogged the economic issues. And yet, if viewed from all sides, the development of this interrelated group of Southern resources, efficiently or badly, is the most crucial article on the agenda of the South's plans for industrialization.

13

Gas and Oil

WORLD WAR II pitchforked the petroleum companies into the chemical business. They are there to stay. So the day may come, not too far off, when you will drive up to your pet filling station and buy a box of aspirin tablets, a set of plastic window panes, or a new spare tire—all of them made by the producer of your favorite gasoline.

For the South this raises critical questions. Since crude oil and natural gas have become raw materials of a new and greater chemical industry, will that industry establish itself in the South where most of these rich supplies originate? Or shall these continue to be piped away to enrich the overcrowded industrial areas of the North and East?

The petroleum companies do not plan to go into the chemical business. But they are surely on that road. Their refineries turned out incredible quantities of three vital war munitions as chemical as sulfuric acid: aviation gas, butadiene, and toluene.

Aviation gasoline is a tailormade synthetic fuel, not

found in any crude oil on earth. Butadiene is the single most important raw material for synthetic rubber. Toluene for the explosive TNT is now chemically twisted out of petroleum whereas during World War I we knew only how to sweat it out of coal tar. These same three are raw materials not only for making synthetic fuels and lubricants, antifreeze, solvents, and rubber, but also plastics, fibers, waxes, lacquers, and all the dyes, perfumes, flavors, and medicines synthesized from coal tar.

May 25, 1942, is a red-letter day in the annals of Baton Rouge. Though that city is the capital of Louisiana, the date is not a legal holiday. In fact, it means nothing in the state's tumultuous political history. But no one who was at the big Standard Oil refinery will ever forget that warm, sunny Monday which opened a new chemical age for Southern industry.

It is always a thrilling moment when the valve is turned that starts a new chemical operation. It might explode and send somebody off in an ambulance. It may go dead and dash high hopes to the ground, to say nothing of wasting months of overtime and thousands of dollars. So everyone from the plant manager to the bottle boy in the laboratory knocks off and gathers 'round.

This world première at Baton Rouge was packed tight with extra drama. The new, simplified process of refining petroleum on trial would, it was hoped, enormously increase our output of aviation gas and synthetic rubber, both vital war munitions that May 1942.

From the top of the two-hundred-foot steel tower to the ground, every inch of the elaborate apparatus had been checked and double-checked. On a platform one hundred and fifty feet in the air, telephone in hand, stood Henry Voorhees, the technical boss. On the ground, Marion Boyer, the plant manager, nodded the go-ahead. The valves were opened. A brief moment of heart-stopping suspense, and nothing happened!

It did not work. But the trouble was soon diagnosed. The temperature in the reactor chamber was too low.

Someone suggested building a fire in the reactor, to "heat her up." It was a scary proposal. What if it blew off the top and ruined a million dollars' worth of apparatus? Boyer made the decision and five gallons of torch oil were gingerly poured in, then five more, then a barrel. The reaction started!

"Sweet Lady Luck put her arms around our necks and kissed us on both cheeks that day," said Marion Boyer afterward.

All over America petroleum men had been watching, and in Washington and London, war-strained generals and admirals anxiously awaited the good word that this new plant had come successfully into production. The hard-pressed Royal Air Force was calling for more, more, always more aviation gasoline. Our own great streams of fighters and bombers were just beginning to roll off the assembly lines. Still more of the extra-premium fuel would be needed if they were to take to the air against the enemy, and the Allied strategists had agreed that air supremacy was the key to victory over the Axis.

More than all this, our synthetic rubber program was at a highly critical point. Half of our desperately needed supplies depended upon butadiene, and this new process held the best hope of importantly increasing this necessary rubber ingredient.

Two weeks later the new Baton Rouge fluid "cat-cracking" plant was turning out half again as many gallons as it was hoped it might.

Petroleum refining has become a very complicated operation, involved with chemistry and bristling with uncouth technical terms. Therefore we can enjoy the full flavor of this meaty, new development more fully and relish its meaning in future Southern industrial development, if we fade out from that red-letter day at Baton Rouge to Pittsburgh in 1857. The million-dollar labyrinth of platinum and stainless steel which is the modern refinery becomes a little iron pot with a long copper neck in which a canny Scot, Samuel Kier, first refined crude petroleum.

Kier's refinery was built on the same principle, and was as simple as a moonshiner's still. He fed five gallons of crude oil into his pot. Heat from the coal fire beneath drove off the volatile kerosene, which condensed again as a liquid and was collected.

James Curtis Booth, the consulting chemist whose advice put Kier in the kerosene business, had found that when he distilled crude petroleum carefully in laboratory apparatus, different products separated off at different temperatures. But Kier was interested in only kerosene. This was his sole product, and in those free and easy days he dumped the residue into the river.

Petroleum is a mixture, a terrifically complex mixture, of literally thousands of different gases, liquids, and solids. Yet all these are composed of only two kinds of atoms, carbon and hydrogen. These two are joined together in a staggering number of combinations to make a multitude of compounds called hydrocarbons.

Now hydrocarbons are not haphazard combinations. They are all constructed upon a marvelously simple scheme, as plain and regular as the cells of the honeycomb. If you follow through the simplest of these compounds you see that they come in regular series of so many hydrogen atoms (H) with so many carbon atoms (C), the little number behind each telling how many atoms of it are in this compound. The most familiar hydrocarbons found in crude oil are the paraffins, of which methane (CH_4), ethane (C_2H_6), and propane (C_3H_8) are the simplest members. They all contain twice as many H atoms, plus two, as C's, so the chemist lumps them under the general nickname formula,

C_nH_{2n+2}. By breaking away those two extra hydrogens, he gets another important family of hydrocarbons, the oleofins, which may be expressed by the symbol C_nH_{2n}. Ethylene, propylene, and butylene are the three lowest members of this group, which is not commonly found in crude oil as it comes from the earth, and the last of these is a very important ingredient of synthetic rubber.

In the horse-and-buggy days, when petroleum refining was simple distillation, the product was kerosene, with lubricating oils as a sideline. The ten per cent of gasoline that was recovered willy-nilly was a bothersome waste. Petroleum men then cared and knew as little about the hydrocarbons in crude oil as they did about the Aztec gods.

The electric light and the horseless carriage changed all that. By 1910 the petroleum refiners were out of the illuminating business and in fuel. The kerosene lamp was relegated to the attic, while the demand for gasoline grew like a pumpkin vine on an August afternoon. Half a million sputtering automobiles were jolting over the rutty roads that year. During the next decade they were to double every twelve months.

Having stretched to the limit the gasoline that distills out of petroleum, the refiners began eking out with gasoline squeezed by pressure out of natural gas. Even before Pearl Harbor, sixty million barrels of this casing-head gasoline were being recovered.

And still the demand for gasoline grew and grew.

In despair the refiners turned to the chemists for help. John D. Rockefeller was one of the first petroleum men to suspect that there might be a good deal of chemistry involved in this business. He installed a real research department and among the chemists he hired was a strapping young Clevelander with a Ph.D. from Johns Hopkins, William M. Burton.

Burton worked out a practical plant operation of breaking or "cracking" the higher carbon-hydrogen combinations down to the lower gasoline range. This cracking process just about doubled the output of gasoline from a given crude oil. It also demonstrated that chemistry is a bread-and-butter proposition in the petroleum business, and soon cracking by heat and pressure graduated to chemical processing.

Cracking induced by means of catalysts, those mysterious substances which hep up chemical reactions but themselves remain inviolate, was next found to give greater and better quality yields of petroleum products. It produced a gasoline of higher octane value and more susceptible to further improvement by tetraethyl lead. Our powerful war weapon, aviation gasoline, therefore, is a product of "cat-cracking," which is the petroleum man's slang for the catalytic process of cracking crude oil, first developed by Houdry.

But this vital new process, perfected shortly before the Nazis marched into Poland, has a serious operating drawback. The pellets of catalyst through which the petroleum vapors pass is quickly fouled by a sooty coating of coke. To restore its efficiency, the catalyst surface has to be periodically cleaned by hot air, so hot it literally burns the coke off. This rejuvenation of the catalyst means either stopping the flow of crude-oil vapor through the reactor or removing the catalyst for cleansing.

That memorable day in May, 1942, Baton Rouge tested out an answer to this problem. In its simplified cat-cracking process there are no catalyst pellets, rather the catalyst is introduced direct into the oil stream in the form of a powder quite like coarse talcum. The catalyst comes in fresh and white at the bottom of the gigantic reactor; it exits in a black cloud at the top, where it is collected, rejuvenated, and recycled back, a continuous circle. To perfect "fluid" cat-cracking, as it is called because the catalyst flows through the system like a liquid, took six years.

Today there are thirty-five cat-cracking operations the world over; the Baton Rouge refinery is turning out more "100-octane" gasoline than it did ordinary gasoline before the war; and tests prove that cat-cracked gasoline has superior antiknock properties.

Aviation gasoline is a chemically manufactured product, a fuel as synthetic as a sulfa drug or a fingernail lacquer. The chemists are not content to unscramble the omelette of hydrocarbons in crude oil. They must separate yolks and whites of the individual eggs and then concoct an entirely new soufflé, a blend of ingredients both above and below 100-octane. This is a very delicate business, full of superfine details discovered sometimes after months of precise research, sometimes by lucky chance.

At the Phillips Laboratories in Bartlesville, a young chemist was making routine tests, adding definite quantities of tetraethyl lead to gasoline samples and checking the antiknock results. One sample failed to respond. He repeated the experiment with the same disconcerting result. Puzzled and fearing he had made a mistake, he went to his supervisor and together they rechecked. The youngster had not slipped. Analysis of the gasoline sample revealed a minute trace of sulfur, and cross-checking by scores of petroleum chemists established the conclusion that the sulfur in gasoline gums up the lead treatment.

"And the devil of it is," explained Ross Thomas, in charge of Phillips' chemical developments, "it pays more to take out the last one-hundredth per cent than any other. But it also costs more and takes more pains."

Thus aviation gas has been created out of a myriad of such scientific minutiae, each one of which had to be recognized, interpreted, and then applied. An outsider can hardly imagine the details, but the result is plain to all. From a few gallons available for some test flights in 1936, the output has grown to more than half a million barrels—that is over two hundred and ten million gallons—a day.

We have suddenly become an octane-conscious people. The protesting ping, ping, ping of the old car when we stepped on the gas proved to our dissatisfaction that the rationed fuel doled out during war days was not up to good, old, prewar stock. Newsreels showing great flocks of mammoth bombers springing into the air like a bevy of quail were an impressive demonstration of the power of aviation gasoline. We are all quite ready to believe that 100-octane gas is a miracle fluid that will run the old family jalopy from here to there and back again on a single gallon.

But we are riding for a fall unless we realize that we shall get a bad bargain at any price if we put 100-octane gas in the old car. It is as wasteful to burn high-test gas in a low-compression engine as it will be silly to drive the supercar of tomorrow on low-test fuel.

Assured on the highest authority that we are using gasoline faster than we are finding oil, we have flopped from the cheerful assumption that we have petroleum for a thousand years to the sneaking suspicion that some day, maybe pretty soon, the skeleton of our wasted natural resources will creep out of our national closet and lock the garage door. This is a very serious matter to a people which would rather mortgage the homestead than give up the car.

"Sure," the pessimist says, "better, cheaper synthetic

tires are fine and so are new plastics, but come now, Brother Chemist, be reasonable. If you are going to make a million tons of rubber and start building cars and planes and the promised homes of the future out of your new plastics, what about gas for my car and oil for my burner?"

The chemists have their answers ready. Constitutionally they are an optimistic breed and having so many times duplicated nature's products—and often improved them markedly—they have their own ideas about natural raw materials. Twenty years ago, one of the greatest of them, the late John Teeple, told an international conference on world resources: "The silliest thing we can do is to save anything for our grandchildren. The chances are 25-to-1 they will not want it."

Almost instinctively chemists believe this is a sound principle. Future generations, they point out, will not need many materials we consider critical, for the simple reasons that either they will have stopped using them or found better ones. Teeple cited chapter and verse of the great wail our own grandfathers raised over the vanishing supply of whale oil for their lamps. Leviathan was being harpooned to extinction—just before kerosene was successfully distilled out of Pennsylvania crude oil! Good scientists are like Teeple: their beliefs are well buttressed with facts.

Since the chemists, the geologists, the engineers of the petroleum field refuse to become panicky over the prospect of oil depletion, let us take a look at the reasons for their confidence. At the outset they face the fact that we will use much more liquid fuel. The airplane and the Diesel engine will take care of that. But aside from the off chance of the discovery of an entirely new engine to use a new type of energy—atomic energy, for instance—they point out that there are many sources of liquid hydrocarbons suitable for fuels. Here petroleum's partner, natural gas, bobs up serenely.

For twenty years natural gas has been elbowing its way forward as a supplementary source of petroleum products. Before the war it was supplying almost a tenth of our gasoline in the so-called casinghead gas squeezed out by pressure. This up-and-coming junior member is now ready to take over from twenty to twenty-five per cent of the firm's motor and aviation fuel business and to assume active management of the new chemical department.

Natural gas started by supplementing gasoline supplies. It went on and established its own business in bottled gas, fuels made from its butane and propane of the general type of Pyrofax and Philgas. It branched out into strictly chemical enterprises under the auspices of Union Carbide and Carbon Corporation and built up a brand new, distinctively American industry, producing such unexpected products as alcohol and antifreeze; plastics and fibers.

During World War II natural gas took over other

new jobs, supplying big tonnages of essential chemicals for aviation gas, synthetic rubber, and high explosives. Thanks to these illustrious war records, the people in the gas and oil sections of the South and Southwest have learned of the chemical values that are locked in both these raw materials.

Especially throughout the Gulf Coast region they expect these chemical values to create new, profitable industries. This expectation has turned the old issue of the conservation of these natural resources sharply to the right. No longer is it a worthy cause supported by starry-eyed enthusiasts. It is a practical business and political issue—will Gulf Coast gas and oil be held and processed right where it is produced for the benefit of the home folks?

Naturally, enthusiasm in the South for the three famous war-built pipelines that drain the Southwest fields is very chilly. The Little Big Inch carrying gasoline refined in Texas City to New York is tolerable. But the Big Inch delivering three hundred thousand barrels of Texas crude oil every day to refineries in the Philadelphia-New York area is a very different matter. As for the twenty-four-inch pipeline of the Tennessee Gas and Transmission Company, which takes two hundred and seven million cubic feet of gas one thousand one hundred and sixty-five miles from Corpus Christi, Texas, to Charleston, West Virginia—well, the mere mention of it gives a lot of Texans heartburn.

What causes the local patriots special distress is that this highly potential Texas raw material is being used for fuel within spitting distance, as they say, of the great anthracite coal fields. To them, that just does not make sense. Their hot resentment has been fanned by hearings held last winter by the Texas Railroad Commission, which is charged with oil and gas conservation.

The testimony that roused the Texans involved a number of well-known names and some important federal agencies. The Tennessee Gas & Transmission Company was originally formed by a syndicate in which Curtiss Dall, former son-in-law of President Roosevelt, was an active member. Its original purpose was for a pipeline from the Opelousas field. When this was vigorously protested by the state of Louisiana, the Federal Power Commission permitted the company to switch its application to Driscoll in the South Texas fields. Notice of that revised application was "lost" in the office of Texas Railroad Commissioner Ernest O. Thompson. Accordingly, Texas was not represented at the Washington hearing on this revised application, and the Federal Power Commission rushed through an authorizing license in a record twenty-one days. Up to the granting of the license apparently nobody in Texas knew anything about this proposed Texas-West Virginia line. A loan of $44,000,000 was secured from the Reconstruction Finance Corporation, which is said to have been the only real money invested in building this pipeline. On the plea of war necessity, A-1 priorities were secured for critical materials. On the same excuse, the vigorous protests of Texas Governor Stephenson and Railroad Commissioners Olin Culbertson and Beaufort Jester were brushed aside in Washington. To add insult to injury, Texans learned that Texas gas is being sold cheaper to small consumers a thousand miles away and in the heart of the coal country than it is in the Lone Star State.

This wad of ammunition has been used by Texas newspapers to spray the countryside with incendiary bombs well packed with denunciation of leading governmental figures. The political implications are plain. The conflagration has spread to Louisiana, to Mississippi, to Florida, and the people have been getting a lively, understandable course in gas and oil economics and the chemistry of the hydrocarbons. Egged on by all these popular agitations, the legislators of these states have been scratching their heads to devise some legal means of making it impossible, or at least unprofitable, to carry these raw materials away from the South for processing in distant states.

Meanwhile the technical men who are creating new chemical values out of these Southern raw materials go on working and say little. But, as Dr. Schoch, of the University of Texas, has said: "if five cents per thousand cubic feet can be added to the cost of gas at the well mouth—its uneconomic uses as an industrial or household fuel in states near the coal fields are eliminated, and gas becomes a raw material for gasoline, plastics, and scores of medicines." By this procedure, Dr. Schoch adds, "Waste of our natural gas resources, long a national scandal, will then be stopped more quickly and more completely than could be done by a bookful of laws."

That is not disparaging all that has been accomplished by Louisiana and Texas legislation to curb these wastes. Today the gas industry stands where the oil industry stood twelve or fifteen years ago. But sane men do not light cigarettes with five-dollar bills, and the gas men are learning to like sensible regulation just as the oil men now gladly cooperate in compulsory proration, which they had to take over their violent protests.

Thus the new chemical values in oil and gas are active allies of conservation, helping importantly at the very source to save our reserves. More than this, when these materials are employed in chemical processes every last drop of usefulness is wrung out of them. Converting wastes to values is the industrial chemists' favorite occupation. Another pet job of theirs is finding substitutes for materials that are getting scarce and too high priced.

Ersatz gasoline from natural gas is already a fact and it is not the sole supplementary resource that we can tap if our petroleum supply fails. A chemical's-eye view of the fuel future was focused for me by two-score petroleum technical men from Tulsa to Port Arthur, from Baton Rouge to Brownsville. It looks like this:

"Periodically we have been warned of our failing petroleum reserves, which have recently been said to be good for fourteen years. Such estimates have always been revised to more distant dates by the discovery of new fields. Since fourteen years was set, oil in quantity has been found west of the Pecos River in Texas and east of the Mississippi all the way around to Florida. There are surely big, unexplored oil fields offshore all along the Gulf of Mexico.

"Not only is there more oil in reserve, but through deeper drilling and better conservation, existing fields will yield more oil than in the past. Joint improvements in fuel and engines will also stretch our petroleum supplies.

"Eventually we shall assuredly run short of petroleum. However, we know this and we shall import more crude, first from Central and South America, later from the Near East.

"The first sign of serious depletion will be rising prices, and the price of fuel oil, because of increased industrial and marine demands, may soon rise above the reach of the householder for home heating.

"Higher prices for all petroleum products will be a compelling incentive to more efficient refining. Twenty-five years ago we got a quarter of a barrel of gasoline from a barrel of crude. Now we get a half or better. We know several processes, today uneconomic, which could increase the fuel yield, if their operation was a little more efficient or the price of gas a few cents higher.

"Gasoline from natural gas by the Synthol process already appears to be practical and economic. It might well increase our fuel output by a fifth. Natural gas reserves are calculated at 110 trillion cubic feet and we are using approximately 4 trillion feet a year. On the basis of equivalent fuel-value, gas reserves at least equal our oil reserves.

"When we exhaust both oil and gas, Western shale deposits are sufficient for our normal gasoline consumption for sixty-five years. Domestic 'tar sands' contain enough oil for gasoline for a century. In the last extremity, half of the coal in the world lies beneath the North American continent, sufficient to give us all the solid and liquid fuels that we need for one thousand years. Processes are known to convert these three raw materials into liquid fuels."

The chemists assure us that the manufacture of chemicals from the hydrocarbons in petroleum will not jeopardize our gasoline or lubricants. "All the organic chemicals made during 1944 in the United States—and that includes plastics, fibers, dyes, medicines, solvents, acids, chemicals, everything made from all raw materials—

totaled only seventy-eight thousand barrels a day. Daily production of crude petroleum was four million five hundred thousand barrels. Chemicals are literally a drop in the oil bucket, less than three per cent of our gasoline. As for synthetic rubber, we could produce a prewar year's natural rubber imports from two days' supply of crude oil. Besides, don't forget that many of the hydrocarbons most useful for chemicals are useless for gasoline and lubricants."

These are indeed comforting assurances. But best of all, this confident chemical view of the gasoline future does not lull our petroleum people to sleep. The finger of fate, or better, of science beckons them to a new career. They are off on the new job as eagerly as a boy to a ball game.

That new job is chemical synthesis, making from gas and oil not only gasoline and lubricants, but great families of consumer goods, rubber, plastics, fibers, lacquers, perfumes and flavors, gums and waxes, insecticides and disinfectants, medicines. As certain as sunlight these new products will become better and better, cheaper and cheaper. Synthetic products always do.

Chemical synthesis from gas and oil will be on a scale hitherto unknown. From these new chemical raw materials of the South will spring vast new industries, making not only these synthetic products, but processing and fabricating them into all manner of wares from huge tires for trucks and planes to stockings and floor polish and baby rattles. These new industries will build new plants, create new employment, distribute new wealth. Built on Southern raw materials should they not be located in the South where the oil and gas are found?

14

Chemical Treasure Trove

BY PRECEPT and example Doctor A. P. Beutel furnishes one perfect answer to that question of Southern chemical raw materials processed in the South. He was born in Ohio. In Michigan, starting at a draftsman's desk, he climbed up the ladder to general manager. But during the past five years he has become as Texan as a barbecued pig.

"Dutch" Beutel—almost nobody calls him "Doctor Beutel" and even good friends do not know what his initials "A. P." stand for—is in charge of the $90,000,000 colony of the Dow Chemical Company at the mouth of the Brazos River. Here magnesium, the lightest of structural metals, is extracted from sea water, and styrene is turned out in vast quantity for synthetic rubber and plastics. In a third new plant a number of Dow's other chemicals are made from Texas salt, sulfur, and petroleum. The general manager of this, the biggest chemical operation in all the South, is a plain, honest, outspoken engineer. "Dutch" is as forthright as a sledgehammer. He is also a shining example.

I had been warned of the miraculous transformation in this old friend of mine. So, when I found him entrenched behind his horseshoe desk that is built into his modernistic, air-conditioned office, I hailed him with a raucous request, "Let me see your feet."

"Hell's bells!" he exploded, "you've been talking to that dizzy bunch in the New York office. "No," he went on with slow emphasis, showing me his shoes, "I do *not* wear cowboy boots to the office. But," he added with a grin, "I do wear them Sundays."

"Dutch" is inordinately proud that over the week end he is entitled to wear both his trouser bottoms tucked into the tops of said boots. According to the niceties of range etiquette this means that he has a thousand head of steers on his ranch.

"I sure do like it down here," he went on in the true Tajos vernacular, "and do I like running a chemical plant in the South! Why, I wouldn't go back to that frozen Michigan, not for—" He broke off, not able to name a price that did not sound too ridiculous.

Right here Doctor Beutel becomes a symbolic figure. His pro-South ideas are shared by hundreds of Northern chemical operating men who have been playing headline roles in the chemicalization of the South. These ideas are grounded on dollars and good sense. They are the good reasons why three out of four of the four billion dollars put into new chemical plants during the past four years has been spent south of the Potomac.

For war needs or for postwar expansion plans this was one point upon which Army and Navy, Government officials and industrial executives, all agreed. The American chemical industry is moving South.

Down South a chemical plant can be built in the open. It needs no weatherproof building; even a tent-like roof is often quite unnecessary. Its stills and autoclaves, joined by a network of pipes, need no thick layers of insulation. If gas or liquid is to be carried from one unit to another, there is no need to forestall freezing by burying the pipes deep below the frostline.

At Baton Rouge I was discussing these climatic advantages with Marion Boyer, who immediately staged a neat demonstration. From the window of his office he pointed to the neighboring plant where du Pont makes tetraethyl lead.

"See that fine building down by the river," he said. "That is their original unit. We told them they did not have to build a fine university hall like that. But no, they knew how to build a chemical plant, and they did. They learned quickly enough. Just look at the rest of the property."

Like any Southern oil refinery, it is spotted with apparatus all exposed to the elements. The saving is considerable—more than half of the original investment in construction, and "Dutch" Beutel estimates that the savings in time, labor, and materials for repairs and current maintenance are a third. Some of his contemporaries say that this is high, since he is comparing Texas with northern Michigan where at both ends conditions are extreme, but at least the saving is substantial.

These climatic advantages are enticing and profitable. They make a persuasive appeal to those who know that in the big half-moon from Jacksonville to Corpus Christi is the most complete collection of chemical raw materials on earth. Over a century ago a brilliant Irishman, who was the father of the English chemical industry, said, "The foundations of my business rest on salt, lime, and sulfur, and the greatest of these is sulfur." No other region combines these three basic inorganic chemical materials in such abundance.

Sulfur, melted underground with hot water, is pumped in golden streams from half a dozen great deposits scattered from the marshy bayous of the lower Mississippi Delta all along the flat coastal plains of Texas. Beneath the whole Gulf Coast lie colossal beds of pure rocksalt. Time and again wildcatters for oil have bored for a thousand feet into these beds, and disgusted, drawn their drills still in the salt. Limestone is barged down the Mississippi from Alabama, or better, the Gulf Coast is fringed with great reefs of oystershells, beds of almost pure calcium carbonate, to be dredged up in two-ton clamshell buckets.

Close by in New Mexico is potash. In Florida, Tennessee, and South Carolina is phosphate rock. Both are raw materials for fertilizers and for an increasing number of important chemical products. There is mercury in the Big Bend of Texas; bauxite for alum in Arkansas or across the Gulf of Mexico from British Guiana; fine clays and ochres in Georgia and Alabama; native gypsum and fine silica in many locations. This sector of the South has almost everything, and what is lacking can come in by barge or boat, the cheapest of all transportation. This last asset completes the economic combination. Most chemical raw materials are bulky or heavy and low priced. They cannot afford high freight that can be absorbed by finished products.

The South has also a vast store of organic materials needed in the synthesis of textile fibers, plastics and resins, waxes and glues, perfumes and medicines. Here are to be had two forms of cellulose, from cotton linters and from pine wood, for manufacturing paper, transparent wrapping sheets, photographic film, smokeless powder, the finest of lacquers. At hand, too, are vegetable oils from cottonseed and peanuts; proteins from cottonseed meal, soybeans, and grain sorghums; carbohydrates from sweet potato or rice starch or the sugar cane of Louisiana. Southern rosin and turpentine are already the base of exciting chemical developments. Lignin, tall oil, and furfural, recovered wastes that have chemical futures, are all available.

Most important of all, petroleum and natural gas, perfectly able to replace coal, coke, and coal tar, both as fuel (chemical manufacturing needs lots of power and heat) and as chemical raw materials, are here in bountiful supply. They are undoubtedly the strongest magnets drawing chemical enterprises to the Gulf Coast area. And quite recently this attraction has been made stronger.

A little cousin of the atom-smashing technique behind the atomic bomb has appeared in Texas. For the maker of chemicals it packs similar revolutionary ideas. That same Professor Schoch who wants to see chemical values established for natural gas has been subjecting hydrocarbon gases to electrical discharges. He has perfected a process for turning methane into acetylene which may blast the way to an entirely new kind of chemical operation.

Some chemicals react spontaneously. A bit of zinc metal dropped into sulfuric acid fizzes like an opened bottle of soda. Heat promotes reaction among more stubborn molecules. Changes in pressure, greater or less, also assist. For years these have been standard tools of the chemical maker. More recently, catalysts, which stimulate reactions but do not enter into them, have been very useful, and an electric current passed through a liquid or solution is the base of the flourishing electrochemical industry. Bombarding gases with electrical charges is the newest idea in molecule making.

In two of the big petroleum companies' laboratories, men are rigging up tubes of various sizes and shapes, filling them with different hydrocarbon gases, pelting them with different frequencies or voltages. It is a game of blindman's buff. As yet they cannot foretell the results. The men doing this work are as excited as a terrier at a woodchuck hole, digging frantically, but doing very little barking.

"Honestly, we don't know anything yet," protested one of the heads of these researches, "but we do believe that we have a big idea by the throat."

He went on to explain, "The atom isn't simply a speck of indivisible matter. It is an infinitely small electrical system at the center of which is a positively charged nucleus around which a varying number of negatively charged electrons revolve at dizzy speed. The molecules are combinations of these electrical systems. Attacked by electricity, they break up into new combinations. We don't know why or wherefore; we can't foretell results. A vast amount of pure research in subatomic chemistry must be completed before the new textbooks can be written, but we are playing with a new way of inducing chemical reactions that I believe will lead us to a brand-new type of chemical manufacturing."

As a practical chemical-making tool this stunning new technique is still largely in the future. But a new chemical raw material is today abundantly available in the South. Its implications are as astonishing as the new technique. Its application is already a dollars-and-cents fact.

As the exploring drills have bored deeper and deeper

into the earth's crust, a very heavy, very light-colored oily liquid and a colorless gas have gushed forth. There is nothing new in this underground alliance of gas and oil, but this gas and oil are different from ordinary natural gas and petroleum. In fact, they are so distinctive that conservation commissioners had trouble making up their minds whether to rate these operations as petroleum or gas wells. The drillers call them "gasoline wells," the chemists, "gas-condensates." You will hear a lot about these gas-condensates in the near future, for they are going to do things to high-octane gasoline and give us some new Southern industries.

This new type of gas-oil has been discovered all around the coastal region from Alabama to Texas, in Arkansas and California, in the Rocky Mountains, and up in the Turner Valley field of Canada. In other words, gas-condensates are not a local phenomenon, and it is fascinating, indeed, to speculate how they are formed deep within the earth where the temperature is high and the pressure reaches about a thousand pounds to the square inch.

When compressed, most gases become liquid. Upon this principle liquefied hydrocarbon gases are delivered to you in stout cylinders as "bottled gas" for the kitchen range. Liquefied carbonic acid gas supplies the fizz for the soda fountains. Within the earth natural gas is maintained in this liquefied state by the tremendous pressure. When an underground reservoir is tapped, it rushes to the surface, where, the pressure being removed, it escapes as a gas. But still deeper within the earth, where the pressure becomes as high as a thousand pounds, then the liquefied natural gas again becomes a gas, or, as the chemists say, it re-enters the vapor phase.

When such a deeper reservoir is tapped, the liquefied gas reappears in the petroleum mixture only at the point when just the right combination of pressure and temperature is reached. We cannot control the underground temperature but we can maintain the pressure above that saturation point which holds the hydrocarbons in the gaseous state.

For twenty years gas has been returned to oil wells to maintain the underground pressure so that as much petroleum as possible can be brought from the reservoir to the surface. Cycling a gas-condensate is not quite so simple. The rich gases from underground are collected and are literally scrubbed with an absorption oil which sops up the heavier hydrocarbons which are then removed by distilling. The denuded oil goes back to the absorbers while the stripped dry gas is compressed again and returned to the reservoir. The recovered condensate, distilled out of the absorption oil, is then processed by regular refinery methods, to gasoline, kerosene, and heavy oil fractions, and the gases—butane, propane, and ethane—are separated out for enriching fuel gas, or making aviation gasoline, or as chemical raw materials.

This cycling operation is no penny-ante game. Though but six years old, already there are forty such plants, chiefly in Louisiana and coastal Texas. More are being built. Every day nearly four billion cubic feet of gas-condensates are being cycled, recovering some ninety thousand barrels of liquid products which give us a quarter of all our natural gasoline and allied products. No wonder the experts declare that cycling is the most important of all the startling wartime developments in the oil and gas industries.

"Cycling of gas-condensates," says Doctor Frank Dotterweich, "will do more to eke out our gasoline supplies than any discovery since Burton learned to crack crude oil. It will build a dozen new industries in the South and bring the American people enormous wealth in our new synthetic products."

That is a bold but expert opinion. This Johns Hopkins-trained dean of engineering at the Texas College of Arts and Industries is an energetic specialist, keenly alive to work-a-day problems. He resembles the traditional, absent-minded professor as little as Buck Rogers does Casper Milquetoast. Apparently he thrives on hard work. Besides Acting Dean, he holds a double-barreled professorship—gas chemistry and chemical engineering—and he has been on loan as consultant to the Petroleum Administration in Washington.

"We must drill deeper and deeper," Dotterweich reasons, "and as we do, these gas-condensate fields will supply more and more of our distillates and natural gas. In south Arkansas, Louisiana, and the Gulf Coast Plain, they are now half of the new discoveries. Moreover, cycling makes real savings. Naturally, best results obtain with the active cooperation of all the property owners in a given field. In the Cotton Valley of Louisiana, for example, cycling started in 1941 after over one hundred wells had been completed. Through fieldwide utilization, forty wells are producing more than one hundred did. That saves the cost of sixty wells and all their gathering lines. It also assures that we will get every last drop out of that field. Finally, the products of the cycling plant are just what we want, the lighter, more versatile hydrocarbons, perfect for making high-octane gasoline or as chemical raw materials."

This is no empty boast. It is proved to the hilt at the newborn town of Chemcel, Texas, southwest of Corpus Christi, close to the very birthplace of the cycling operation, and a few miles from the college where this gas-condensate prophet teaches chemistry. Here the Celanese Chemical Company, offspring of the company that makes the textile fiber, turns out a whole line of chemicals belonging to the family of acetic acid, which is the acid in vinegar. The enterprise is a neat demonstration of the rapid progress of chemistry.

In grandfather's time the dyers used vinegar as a

mild, impure acetic acid. At the time of World War I, acetic acid came as a joint product with wood alcohol —of evil Prohibition days' reputation—when wood is charred in a tightly closed iron retort. Twenty years ago synthetic acetic acid was first prepared from acetylene, the gas which blazed in the Prest-O-Lite headlights of the Gay Nineties and which today burns hotly with oxygen in the welding torch. Now acetic acid is being made from acetaldehyde cracked out of recycled gas-condensates. It is the same old acetic acid, but the Celanese people think it will be cheaper. If it is, it will cut the costs of rayon and plastics.

Logic as straightforward as a gun barrel backs this new chemical operation. Cellulose plus acetic acid equals cellulose acetate, which is Celanese. The operation is not a war baby, for the process had been worked out at the home plant in Cumberland, Maryland, where an output of ten tons a day had been on schedule. But the new plant in Texas is also turning out thirty-five tons of butadiene a day. That process, too, is a Celanese development, with some novel short-cuts, aimed at lower costs.

Butadiene suggests synthetic rubber, so I asked the manager, Arthur E. Peterson, about the postwar program. He answered in his deep English voice, quite a shock to ears atuned to the soft Texas drawl, and his reply was no less startling than his accent. "The future of butadiene isn't hitched tight to synthetic rubber. It makes some rather fascinating new plastics."

This hints that the new Celanese plant at Guadalajara, Mexico, may not be the only postwar expansion planned by astute Camille Dreyfus for this company. It reminded me of the new low-cost plastic from butylene and sulfur dioxide about which Ross Thomas and Richard Alden, top Phillips petroleum chemists, are so enthusiastic.

"We flirted with these butadiene plastics long before the war," Peterson went on, "and postwar we intend to work up a lot of our chemical by-products. But," he added quickly, "remember that Celanese Chemical is a subsidiary of the Celanese Corporation. Our job will be to supply our parent company with raw materials."

At this point he shied away from a question that is cudgeling many first-class petroleum and chemical brains in these days: "Where do we go from here?"

However tempting the prospects in petro-chemicals, the oil executives declare flatly that they do not intend to get into what they are pleased to call "the chemical gadget business." They say they will make only basic chemical intermediates, raw materials you can ship by the tankcar, leaving to the rubber and chemical manufacturers the job of working these up into finished products. But the rubber and chemical people just cannot help wondering.

It may be true that the petroleum leaders, their eyes

focused on markets for twenty-five billion gallons of gasoline and five hundred million barrels of fuel oil, cannot see such microscopic figures as one hundred thousand pounds of some chemical with a tongue-twisting name. Yet rubber and plastics are not peanuts and popcorn on anybody's sales counter, and these men are not blind to the figures on a balance sheet. The dollar sign says that petroleum companies earn two, four, six, sometimes eight per cent on their net worth, while chemical companies earn ten, twelve, sometimes more than fifteen per cent. That is language any company president understands. And the petroleum companies have the raw materials, the apparatus, the technical staffs to launch forth as chemical manufacturers.

Shell, a virile pioneer in petro-chemicals, already makes a lot of products not sold at their familiar yellow gas stations: acetone, several fancy alcohols, some outlandish-sounding items like methyl-isopropyl ketone and mesityl oxide. Many of these could be quite simply converted to consumer goods. Gulf has long made a flyspray out of its own kerosene. Standard of California markets its mineral oil through the drug trade.

From making synthetic rubber to fabricating tires and tubes is a step that does not need seven-league boots. Both would be "naturals" for sale at the gas stations where already you can buy antifreeze, touch-up finishes, body polishes, radiator cleaners. Even before gas rationing, these sidelines were frequently the rent-payers; during the war they were literally business-savers. To raise the daily sales of the stations is a real problem of the whole industry. Cleaning fluids, household lacquers, and insecticides would slip into the sales groove like a piston in a cylinder, and all can be made from petroleum intermediates. What with pop and candy, sandwiches and soda, fruit, the comics, and picture postcards, these ubiquitous, well-located distribution stations of the petroleum industry may someday soon rival the drugstore and the five-and-ten.

Meanwhile some of these puzzled executives have found a way out of the petro-chemical dilemma by alliances. Phillips Petroleum and Goodrich Rubber started it when they organized the Hycar Chemical Company and long before Pearl Harbor produced a new, strictly American type of Buna synthetic rubber. A wartime version of the same idea is the Port Neches butadiene plant operated by the so-called Port Arthur group of oil companies; Texas, Gulf, Socony-Vacuum, Pure Oil, and Magnolia. Near by at Texas City, the Carbide and Carbon Chemicals Corporation has a working agreement to take ethylene and propane from the adjoining refinery of Pan-American. The newest petrochemical alliance is the Jefferson Chemical Company, owned jointly by American Cyanamid and Texas Oil.

The Jefferson plant is building now next to the Texas refinery at Port Arthur. From both ends of the line, from Frank L. Wallace, works manager of the refinery at

Port Arthur, and from Philip M. Dinkins of Cyanamid, general manager of Jefferson, in New York, I learned the sensible whys and wherefores of this combination of skills and resources.

Said the petroleum man, a veteran of the great refining community in the famous Sabine section, "We have hesitated to employ the by-products of gasoline as chemical raw materials, reaching out into a field where the point of view, the scale of operation, and the selling methods are somewhat foreign to our experience. By joining forces with American Cyanamid we secure all the benefits of their chemical philosophy and experience, while at the same time having a direct share in working up our materials. A petroleum outfit that goes into chemical manufacturing on its own is likely to be first overenthusiastic at the prospect and then overpessimistic when the difficulties of the new technique, the competitive troubles of breaking into a new sales field, and the expenses, pile up."

The chemical man began by recalling that one summer evening fifteen years ago, he and I had cocked our chairs against the porch railing and discussed the chemical possibilities in oil and gas till the moon had set. "So you see," said Phil Dinkins, "the thinking behind the Jefferson plant goes back a long way. But however tempting a new line of chemicals appears, technical progress comes so swiftly and competition is so keen, that you simply must get your raw materials right from primary sources. Cyanamid could find no major oil company willing to sell its by-product gases on the long-term basis essential to justify investment in a petrochemical plant. Our marriage to Texas was inevitable; but, believe me, it was no shot-gun wedding."

More effectively than the wisest State Industrial Commission or the most energetic Chamber of Commerce, such joint chemical enterprises will promote the postwar industrialization of the South. Just as raw materials draw chemical makers Southward, so their products are a magnet attracting other industries.

Air Reduction, producers of hydrogen, and Wesson, refiners of cottonseed oil, have teamed up in the South Texas Cotton Oil Company to make hydrogenated cottonseed oil, which is vegetable lard. At Pensacola, the American Cyanamid Company has a department of its own right in the rosin plant of Newport Industries, where it treats rosin with alkalies from its subsidiary at Corpus Christi to make paper-sizing materials. Next door, U. S. Industrial Chemicals takes in this same rosin, pumped over hot and liquid, and makes synthetic varnish resins. Caustic soda from Diamond Alkali and fats from the stockyards have brought to Dallas a Procter & Gamble soap factory. By reverse reasoning, Reichhold Chemicals has built a new phenol plant at Tuscaloosa because a neighboring paper mill will buy their by-product sodium sulfide.

In the South a lot of new chemicals will presently be

available for many kinds of industries, and much Southern postwar planning is predicated on these materials. Synthetic resins will be made at Memphis, at Tuscaloosa, and at Mobile, and while today there is but one important Southern plastics fabricator making cups and electrical sockets and such like, I was told of enthusiastic plans afoot in Atlanta, New Orleans, and Mobile to enter this promising field. I read the blueprints for at least two big plastic-plywood operations. Both these have been drawn by Southern lumber companies which prewar only shipped "rough stuff" to be finished in Northern planing mills. With all the raw materials now locally available and big market at the front door, new paint and lacquer plants are projected at Houston, Jacksonville, Savannah, Vicksburg, and, I dare say, a dozen other Southern cities.

Most of these projects are fathered by Southern men financed by local capital. In the main the new petrochemical industry will be a big-boy game, played by the strong petroleum, chemical, and rubber companies. But these new "associated enterprises" are the idea of home-town talent. This is a type of Southern industrialization quite distinct from that which depends upon a branch factory of a Northern company bribed by a free plant site or lured by the delusive prospect of a low wage scale. It is at once the promise and pledge of the South's chemical future.

15

Plans for Tomorrow

KIPLINGER'S Washington letters do not often burst forth in prophecy. For all their crisp, gossipy style, these weekly bulletins of backstairs news from the nation's capital, circulated privately among businessmen, are quite circumspect when it comes to dealing with facts and events. Yet under the date of May 26, 1945, their subscribers read these glowing statements:

"The most important economic development in years, except war, occurred during the past week with only scant attention from the public . . .

"It is the change in freight rates by the ICC . . . a step towards industrial decentralization, which will shrink some cities, expand others, rechannel the flow of trade, attract many workers to new locations, make new jobs for others . . .

"Not immediately but five years hence, ten years hence . . . areas west of the Mississippi will gain in new manufacturing, new distribution centers, distribution branches, and new population.

"South and Southwest, the same—a strong stimulus

to them . . . freight differences that have cultivated industry in the East are now in the process of being removed, and the result will be a shifting, an expansion in the West and South, visible within five years."

Vividly, concisely, this quotation summarizes the natural effects of the order of the Interstate Commerce Commission advancing class freight-rates 10 per cent east of the Mississippi and north of the Potomac and Ohio rivers and lowering these same rates proportionally South, Southwest, and West, except on the Pacific Coast. Though the ICC order, in effect August 1, 1945, was as thick as a big city telephone book, it is admittedly but a start. It only affected "class rates" covering groups of goods, almost all manufactured wares, the boxed and packaged items of relatively high value. It did not change "commodity rates" on the heavy bulk stuff such as coal, ore, lumber, steel, grain, gravel, and cement. Revision for the heavy commodities will come. The practical problem is as complicated and intertwined as a tropical jungle, and it will take time to clear away the thickets of interrelated details.

The class-rate order was but the beginning of a thoroughgoing revision of the national freight-rate structure. This has been on the conference table of the ICC since the late thirties. Direct action was jogged by the decision of the Supreme Court to hear the case of the State of Georgia claiming freight-rate discrimination against the South, but the ICC has been thrashing out this question for years and had already made up its mind on the principles involved. This is fortunate for the South. If the Commissioners had been forced to undertake this herculean task by court order and against their will, begrudging, piecemeal changes might have been dragged out for decades.

While sincere equalization may now be expected as soon as is humanly possible and the eventual effects will be all that Kiplinger so boldly sketches, it is asking for disappointment to expect that results will immediately revolutionize Southern industry. The discrimination has been obvious and real. For years, however, it has been noteworthy that most of the agitation for revision has come, not from the big shippers, but from the politicians and the economists, the farm bureaus and the chambers of commerce. Southern industrialists dealt directly with the Southern railroads. Where they could show cause they obtained rate concessions on specific goods to specific points which enabled them to adjust their competitive position. Naturally the equalization of freight rates will help. But unless the ICC abrogates its first duty, Southern industry will not be permitted to enjoy unfairly any decisive advantages.

Lower freight rates will stimulate and strengthen several other influences that will advance the cause of Southern industrialization. The war has been a potent force scattering industry out from the Northeastern zone. For military reasons it was expedient to locate factories beyond the seaboard area that was most vulnerable to airplane attack from Europe. For social and political reasons the Government was well pleased to make the most of this opportunity. The South and West have therefore been spotted with new plants. Many of these were operated as war plants by Eastern companies, and the temptation to curtail operations in the old factories and expand in the new locations with all their up-to-date equipment will be irresistible. New ventures, too, are going to be swayed by this same pull to the South and West.

In the prewar past the bait of low wages and comparative freedom from union interference has been the most tempting lure to bring new factories into the South. These seductive inducements are losing their savor. Wages and working conditions will continue to be equalized and standardized all over the country. The more American industry decentralizes, the more effective will agitation be for union organization in new plants in new territories. There is plenty of popular sentiment to enlist political aid for this cause. Accordingly, in the next few years it may well transpire that the chief benefit of "freight equalization" to Southern industries will be its offset to "labor equalization." In any event, the one inevitable question, raised whenever a Southern plant location has been up for serious consideration, will be swept away by freight-rate revision.

This may often be the final argument that clinches the decision for a Southern location.

Other arguments are sound and increasingly forceful. Climate, for example, is a very persuasive reason, especially to any chemical processing plant. Air conditioning now cancels the obvious counterclaims.

"Up North you buy fuel all winter to keep warm," said a paper mill manager who has worked in Maine and Wisconsin, "and down here you buy fuel to keep cool in summer. Take your choice, for I look to see air conditioning spread so rapidly throughout the South that in a few years you will be no more able to sell a house in Alabama without a cooling unit than you could one without a heating plant in Vermont. And that will go for factories, too."

Southern raw materials also become increasingly attractive as they are increasingly processed on the spot into intermediate products. Plastics, rayon, plywoods, metals and alloys, synthetic rubber, hundreds of chemicals all furnish the basis for whole groups of fabricating industries gathered like bunches of grapes around these new sources of supply. This is apt to be the strongest lodestone of all the attractions pulling manufacturers Southwards. Already its force is felt, and what is happening in Memphis is exciting and quite typical of forward-looking promotional work being done in many centers of the South.

In that historic river city, great marketplace for cotton

and lumber, there settled a few years ago a husky, hustling chemist, Stanley J. Buckman. Born in South Dakota, educated at the University of Minnesota, he moved South after graduation to Louisville where he went to work for the American Creosoting Company. He climbed to the head of the research department and to an outstanding position as an authority in the field of wood preservation. Then he struck out on his own. As a manufacturer of wood-preservative chemicals he did his bit during the war turning out protective treatments for wooden barracks, tent pins, packing boxes, or whatever the Army needed made of wood. The businessmen of Memphis have chosen this practical, successful, technically trained citizen to spearhead the industrial development work of their Chamber of Commerce.

Dr. Buckman resembles the ballyhoo artist of the city-boosting era as little as the modern sales engineer does the old-fashioned drummer. The Industrial Research Committee—significant name, that—which he heads has a program that would bowl over the enthusiasts who labored to identical ends in the distant prewar epoch of the industrialization campaign.

Twenty good men and true, with the help of Charles S. Peete, native-born secretary provided by the Memphis Chamber of Commerce, have been putting Memphis industries through a stiff course in economic self-analysis. Their accumulated experience blankets agriculture, forest products, chemicals, plastics, metals, textiles, and machine products, gas and oil. In the light of the latest scientific discoveries and commercial developments, they are studying these resources of the Memphis area. Their findings, after combined scrutiny, are not broadcast, but are laid on the desks of appropriate company presidents right in Memphis, a practical hint of new or allied products that might be profitable lines for expansion.

They do not prod and they do not preach—in fact, more and more they are called in to check new ideas that the imagination and initiative of Memphis manufacturers have hatched. Neither do they go out into the highways and byways of industry hawking their wares. When somebody comes along with the thought of locating a new enterprise, they have most of the answers ready and know where they can get the others.

The aim of such a program might appear to shoot at the stars. The sights are fixed, however, on a target right in the range of practical reality. The Memphis Industrial Research Committee hold before themselves the definite objective of providing a satisfactory level of local postwar employment by means of sound industrial expansion. This is the nub of the entire Southern reconversion problem, the somber foreground of its brilliant horizons.

During the war no section of the country went through such a turmoil of labor dislocation. From the northern and eastern portions of the South white and black workers alike moved out into the war plants of the East and Midwest. No one has counted that exodus, but in many localities it amounted to a mass movement. In the Deep South the workers flocked—a quarter of a million of them and their families—into the shipyards at Mobile and New Orleans, Panama City, Florida, and Pascagoula, Mississippi. Throughout the whole area the familiar shift from farms and villages to the camps and airfields and the new industrial centers often went countercurrent to the main streams of labor migration.

Amid all the uncertainties of getting these hundreds of thousands back on the peacetime job, two facts appear to be unmistakable. First, many Southern warborn enterprises have little or no chance of peacetime survival. Second, many of the Southern workers will eventually return to their home localities.

The first premise is categorically true. The big explosives plants will undoubtedly be closed down and the big shipyards cannot continue at anything like their wartime scale of operations. Together these were the greatest employers, so that most Southern war workers must seek new jobs. Because of the naturally isolated position of the explosives plants and the inevitable waterfront location of the shipyards, most of them must remove from the scenes of their warwork.

The second fact admits some qualifications. Not all of the workers who have migrated North and East will return to the South, and not all of those who have moved into Southern cities will go back to the farms. To the majority such a homecoming probably holds forth few attractions. Nevertheless, unless they can find employment, a return to the familiar rural community is logical, especially if the Government is willing to pay their way back.

"Nobody starves in the country," was how a big cotton planter of the Delta Country put it, "and if the going gets really tough in the industrial cities, the old farm and the old neighbors and old job will look mighty good. They won't all come back. How many do will depend upon how tough the going gets." In homespun terms that just about summarizes the experienced thinking of the South. At Memphis, and in hundreds of other Southern communities, they face these facts as they see them without blinking. Dr. Buckman and his colleagues are striving, first of all, to anchor their city's war gains. In this they are rather fortunate. A good deal of their recent growth has been along chemical lines, and chemical war products usually have peace uses.

The furfural from corn cobs and cottonseed and rice hulls, produced at the new Quaker Oats plant for the synthetic rubber program, will continue to go into this product, but it can also go into plastics or for petroleum refining. There is a likely supply of phenol, also for plastics, and the Southern Acid & Sulfur Company makes a general line of industrial chemicals. Furthermore, the new Ernst Bischoff plant will continue to make cellulose plastics, and the new Memphis Mayfair Company will fabricate plastics into consumer goods. With cotton linters for cellulose and ground wood for fillers and plywood for laminating, here are all the makings of a thriving plastics community. The recently discovered Mississippi oil fields are providing crude for the new stills of Delta Refining Company, and other new enterprises run the scale from cotton pickers and other farm machinery at International Harvester to frozen strawberries from the new freezing plant of the Braun Packing Company.

Many plans at Memphis—an electrochemical project, pottery and chinaware from west Tennessee clays, cottonseed and soybean oil refining and margarine manufacture, textiles, and machinery—hinge upon low-cost water transportation, a two-way advantage both in tapping raw materials and in reaching consuming markets. The inland waterways of the South are a strong second string to the transportation bow. From Corpus Christi they stretch across the Gulf Coast to Tallahassee, branching out to Houston, Bay City, Lake Charles, New Orleans, and Baton Rouge, then on up the Atlantic Coast through Jacksonville, Savannah, Charleston, Wilmington, and Norfolk to New York. Up the Mississippi Valley they tap St. Louis, Kansas City, Chicago, even Duluth and the Great Lakes. Eastward they reach Cincinnati and Pittsburgh. Via the Alabama and Tennessee rivers, barge shipments can be made to Birmingham and Columbus, Mississippi, to Chattanooga and Knoxville.*

For handling cheap, imperishable raw materials—coal, bauxite, sulfur, clays, lumber—this far-flung system is being used by many localities to an extent that few outsiders realize. Tank-barges carry oils, even acids, in vast quantities. Since it first opened, a dozen years ago, the famous sea-water bromine plant at Wilmington, North Carolina, has shipped the extremely volatile ethylene bromide by barge to Wilmington, Delaware, where it is an essential ingredient in the preparation of the motor fuel additive, ethyl fluid. Sulfur from the lower Mississippi and the Texas coast goes regularly by inland waterways to New York and Philadelphia, to St. Louis, Chicago, and Pittsburgh.

These water highways serve a great sector of the most potential industrial areas of the South. Their usefulness will be greatly increased by advances in barge design won on the landing beaches of Africa, the South Pacific, and Europe. Bigger, better, lighter cargo carriers, operating under their own power and designed for undreamed-of speed, capacity, and convenient loading and unloading, are at hand at salvage prices.

Not only Memphis, but scores of other Southern communities are reckoning with this improved industrial

* See map inside the covers of this book.

adjunct, nor is the "Inland Cotton Capital" the only city where the bigger-and-better boosting has been superseded by sound, factual promotion. "A Statement of Facts"—a title as significant as the name of the Memphis committee—is a study of local resources for chemical industries prepared by the New Orleans Association of Commerce. Its sixty-nine pages contain not a hint of boast or bribe. George Schneider and Sam Fowlkes are admittedly able trade association executives, but the kind of forward-looking work they are doing for New Orleans has rivals. The swing to facts in plans and programs, as well as in the ways and means of economic promotion, is additional evidence of the revolution in Southern thinking.

It is one more of those puzzling Southern paradoxes that these new ideas about industrial promotion should be crystallized into practice in that easy-going, aristocratic, wholly delightful city of Charleston which, for all of its many charms and numerous claims to fame, has not stood out in our minds as a mainspring of progressive action. Nevertheless, this South Carolina city is the birthplace of a professional organization, the Community Research Institute, whose services make these practical ideals of self-analysis and of self-help available for any community.

The idea is as clear and simple as springwater. A group of disinterested experts—civil, mechanical, and chemical engineers, business and farming authorities, each a distinguished authority in his own right—studies a community as it is today. It makes definite recommendations for tomorrow. First, how the usually neglected assets, the existing businesses in manufacturing, farming, and trade, might be strengthened. Second, what new projects offer the fairest chance for local money and local men. Neither of these homegrown betterments affects the third program: to bring in new enterprises.

The Institute fathers the creation of a fund of local capital to aid existing and new projects. This is no slush fund. It is working capital to expand the business opportunities of the community. Its subscribers receive stock and elect trustees who, with the continuing help of the professional experts, invest this capital. The trustees may lend money to a local machine shop that wants to add a nut-and-bolt factory, or they may buy stock in a textile mill that requires additional capital to buy rayon-weaving machines. They might even decide to launch a brand new paint and lacquer factory or to buy land or buildings to be sold on favorable terms to some established business or some newcomer.

Against the old promotional tactics this community corporation has one thundering advantage. It is business. It places community development on a dollars-and-cents basis. It stimulates a lively, lasting interest and it sets up technically guided projects for which somebody is directly responsible to people who have a cash stake in their success. The Community Research Institute employs the clinical method to tap the resources of a group of experienced specialists to formulate realistic programs. It makes no "survey," renders no prettily bound reports that get no farther than the filing case. It implements definite action, encourages profitable self-help.

"In a lifetime of professional practice as a consulting engineer, I have never faced a Southern economic problem, in a going business or in new opportunities, which could not have been solved by our own people to their own advantage."

That statement of combined experience and faith by Frederick H. McDonald, director of this Charleston research group, is the distilled essence of the new thinking that is renovating Southern life and living. After giving me the striking parable of the Southerner and the Yankee who struck gold, which I repeated in the first chapter, this keen analyst continued with this workaday application of these revolutionary ideas.

"Southern economic and cultural advancement must be rooted in more diversified employment and in better income for our people. The process will require a special kind of education. We must apply sound principles and back them up with proofs of home opportunities. When it comes to capitalizing them, I have been finding that unless we can demonstrate commercial operation on at least a pilot-plant scale, Southerners are unwilling to chance investment where unfamiliar technologies are involved.

"Yet, no matter what we do, the future of our industry and our agriculture alike lie in science. This of necessity produces new techniques. It is but natural that new methods and new products are obscure to the layman. He has no familiar yardstick by which to evaluate them. Many of our leaders realize that research is the very essence of our progress; too few appreciate that the mere test-tube predictions of the laboratory will not convince investors. If, however, theory is reduced to the familiar measure of profits in dollars, as demonstrated by even the smallest type of commercial operation, there is plenty of leadership and enough money in the South to carry them on and forward."

16

Highway to the Horizon

EDUCATION and research: clearer, more logical, less emotional thinking, which is (or should be) the purpose of education; research that finds, defines, and then applies facts! It is not only in the South, where Frederick McDonald so clearly sees these needs, that mankind must build the future on these two solid foundations.

In the South as a whole, the foundation of education is much too shallow and wobbly. Sincere Southerners do not attempt to deny this. They advance many explanations for its causes and they also propose many cures. Neither causes nor cures are topics that properly come within the scope of our present inquiry, but there are certain new manifestations of the existing Southern educational system that will very much affect the future development of Southern resources, human and material.

In this very field many of the Southern colleges and universities are doing outstanding work, so outstanding that some critics say their faculty members are better promoters than pedagogues. It is obvious that the traditional absent-minded professor, be he the gentle scholar of letters or the meticulous grubber for facts, is rarely met on any Southern campus. It may be, too, that the alert, businesslike attitude and the workaday thinking in terms of local profit-and-loss which characterize the Southern college professor, particularly the younger man, has more of a political than a professional inspiration. Such criticism degenerates to the captious when it ignores the constructive contributions being made by these hustling experts in the Southern colleges.

Texans enjoy bragging, and if there is a crossroads in the Lone Star State that does not stake a claim to being first in something or other, then I never stopped there to fill my gas tank. Texans are captivated by boasting for its own sake, but if you want the facts and figures of that braggadocio commonwealth you can have them in minute detail and scrupulous honesty. A regular monthly publication of the University of Texas, the *Texas Business Review,* is a model of all factual and statistical studies of the sort. It presents a current picture of the state of industry, trade, and agriculture. It is more than a recorder of facts, for it prints a continuous succession of capital articles by men who write with authority, and restraint, on subjects that range from the theory of liquid assets in a period of inflation to the facts of Texas' share in the synthetic rubber program. This wholly admirable publication, issued under the directorship of Dr. Alonzo Bettis Cox, an internationally famous cotton economist, is a piece of factual research of the first order and an exceedingly practical tool for anyone doing business in the state.

About the easiest idea to sell the president of any state university is a scheme to use its scientific staff and laboratories in any sort of cooperative project with the industries of that state. Variations of this popular idea, all enthusiastically launched and not all successfully navigated, exist from Orono, Maine, to Berkeley, California. South of the Mason and Dixon line both the number and the success of such projects are higher than the national average, and there are two quite different setups at Gainesville, Florida, and Raleigh, North Carolina, which will suffice as conspicuous examples.

At the University of Florida the head of the Engineering Department has pumped new ideas, that square with ideals of better technical education, into this well-worn concept. Joseph Weil has a trained, three-dimension intellect made effective by a friendly, out-giving personality with the courage to back his convictions. In his neat, simple office he impresses you as a broad-gauged educator, just the man to be called in by the Manpower Commission and the U. S. Army to help frame war training programs. In conference with the State Board of Engineers you find him an exceedingly capable professional with practical experience accumulated at Westinghouse and as chief engineer of radio station WRUF. But coax him off bounds, say to the comfortable rocking chairs on the verandah of the Hotel Thomas, and you discover a philosopher with a rare ability to generalize the principles of both education and engineering and to pin down his thinking to pragmatic conclusions.

In both fields Joe Weil and his colleagues in the Engineering Department are furnishing bold, constructive leadership. Individually and collectively they serve as a panel of expert consultants to Florida industries. This, of course, is "old stuff." Their ideals and methods, as summed up by Weil himself, are quite in the modern vein.

"A subsidized industry," he told me, "is by definition unsound. It is just as risky to transplant a Northern factory as to try to grow Northern crops or flowers in Florida. Some will flourish luxuriantly; others simply refuse to grow. In the case of the transplanted industry, the risks and the costs are such that we must be sure we are right before we go ahead.

"Florida among all the states has a truly tropical zone, and we want to capitalize this unique asset. For example, in the tropics the problems of water purification and sewage disposal are distinctive. We have brought to Florida one of the country's great sanitary engineers, retired from the faculty of a Northern university, and his ripe experience is helping us devise a water-sewage system particularly applicable to a tropical city. Because of the differences in the bacteria present and the high temperature, the technique is not only quite different but it is also unexpectedly more simple and less costly.

"As a laboratory for the study of tropical problems—sanitation, medicine, air conditioning, agriculture, even industry—Florida can help the whole United States be a practical good neighbor to the republics south of us. If we can take them workable solutions of their own problems, solved by our skill and knowledge under comparable tropical conditions, we give them something infinitely more acceptable and more valuable than trade treaties or subsidies. Besides we are rendering services that none of our European rivals can offer."

Forecasting the course of hurricanes, draining the Everglades, a score of such specialized problems are under scrutiny, but the purposeful effort to raise the standards of technical education in Florida is probably the most fruitful experiment of this exceptional Engineering Department. A poorly prepared engineer or chemist or geologist cannot hide his deficiencies from his confrères, and professional standards admit no alibi. Accordingly, the weaknesses of Southern education show up most glaringly in these exact branches where the training is definite and the results can be readily judged. A few bright exceptions—law at the University of Virginia, medicine at Duke and Tulane, chemistry at Rice and North Carolina—only throw into deeper shade the general average of mediocrity. Until recently the University of Florida certainly did not stand out in the sunlight. Its new integration with tomorrow's needs is all the more to the credit of President Tigert and his vigorous Dean of Engineering.

"Our aim," to quote Joe Weil, "is not more, but better, engineers. We are making a definite attempt to prepare the student professionally by giving him a well rounded technical education, but not to neglect those other subjects which tend to make him a better citizen with an appreciation of the problems of the modern world."

This protest against specialization is not a novel idea, but again the attack at Gainesville is fresh. A new kind of screening is practiced which Weil likens to the inspection and selection of raw material in the world of manufacturing.

Each incoming student is examined and interviewed. Some are told frankly that it is unlikely they will benefit from a college course and advised not to matriculate. After two years of broad, background instruction and close consultation with a faculty advisor, the students are again sorted to head them into the fields where their interests and aptitudes best equip them for success. This two-years' seasoning period is made the opportunity for a second screening. It marks the end of the general collegiate program, making a definite terminal point at which it is recognized that some men would make a mistake to attempt the stiffer, technical courses ahead. Having completed the general course they are awarded a certificate—an honorable discharge, an easy way out without the stigma of failing to earn a degree. The inadequately prepared, the inept, the incompetent students are thus quite effectively, and almost painlessly, eased out of the classroom. Thus the bugaboo of all tax-supported institutions is charmed away. It is a formula that might work wonders throughout the whole American public education system.

Following a very different pattern the Textile School of the Engineering College at the University of North Carolina is working effectively to the same ends. At Raleigh was an ancient and honorable institution, half a century old, one of the best of its kind, a high-grade technical school teaching the dateless craftsmanship of fiber and fabric. Three years ago it was electrified by one of its own alumni.

R. J. Carter of the class of 1924 did not graduate *cum laude;* but his Carter Fabrics have an A-1 rating. And "Nick" Carter is loyal to his Alma Mater, respected among his competitors, liberally endowed with income and imagination. He went to Frank Porter Graham, president of the State University, and talked common sense in this fashion:

"A school can't be better than its teachers. To attract good men you must pay good salaries. What if I went out and raised $50,000 or $100,000, or whatever we need to get the best talent in the world for our Textile School?"

The answer to that question is the North Carolina Textile Foundation which has raised over $700,000. This tidy sum has been contributed mostly from the textile mills of the state and every penny of it is pledged to be spent for men. The Foundation is now out to raise an even million.

At the wheel of this practically minded project is a real textile executive, Malcolm Campbell, as canny as his name. He is quick and decisive as an adding machine, but the spark of Isaiah glints in his brown eyes when he begins to talk about what lies ahead for us all in the field of fine fabrics.

In building up the manpower of the North Carolina Textile School, one of Campbell's first acquisitions was Elliot Grover from the Massachusetts Institute of Technology with a business record of successful management of one of Rhode Island's biggest mills. The latest is Dr. Frederick T. Pierce, for twenty-three years at the famous English school at Shirley, who will head the new department of textile research.

To train young men and women superlatively well for careers in the textile industry is the first object at Raleigh. The second is to give expert assistance to the textile mills of North Carolina where more cotton spindles and more rayon weaving machines are located than in any other state. Beneath all lies an ambitious program of research in which knitting and synthetic fibers are prominent, for the North Carolina industry, formerly devoted almost exclusively to cotton goods, is reaching out into new lines and new fabrics, such specialties as shoestrings and rugs, fine fabrics made of cotton mixed with all the new synthetic fibers.

This industry-supported Foundation is additional evidence of the new self-help for cotton in the Southern textile industry. From one of the most ancient arts, the making of cloth has grown by traditional, rule o' thumb methods, to be one of the greatest of modern industries, almost without the aid of science. During the war there has been a great awakening and, as Malcolm Campbell says, "We have shaken the lead out of our shoes and are going places."

To say nothing of the research work of the National Cotton Council, the textile men have their own nationwide, all-fiber research under the Textile Research Institute with headquarters at New York and working outposts at Princeton and the University of North Carolina. At Charlottesville, Virginia, is the Institute of Textile Technology, with ultrascientific aims, backed liberally by millowners and led by Ward Dulaney, an atomic bomb of ideas and enthusiasm, formerly with the Institute of Paper Chemistry. At La Grange, Georgia, the Calloway Institute, financed chiefly by Fuller Calloway, aims to do specific research for mills, specializing in the conservation of assets by doing better jobs with existing machines. There is a new, highly technical magazine, *The Textile Research Journal,* and textile research directors now sit down together at the meetings of the Industrial Fibers Institute. Under the chairmanship of John Elting of the Kendall Mills they toss back and forth questions and answers.

In spite of the clamps screwed down by war conditions, out of all this yeasty ferment of research have already come new materials and new methods. Good worsteds can be made now on cotton machinery out of wool tops or wool and rayon. The saving is a quarter of the old costs. New machinery can handle fiber lengths up to three inches. A new flat knitting, called tricot, yields a big production of amazingly beautiful fabrics. Spinning and weaving are wholly eliminated in a new process by spreading out oriented fibers, then impregnating them with a binder, and applying heat to produce a disposable cloth at low cost for table and bed "linens" which have long been cottons.

The Navy approved such a disposable white collar for officers. They are made of layers of cotton-paper-cotton, formed in great sheets. The collars are stamped out complete with seams and button holes. Wear them a day, turn, and wear another day, then toss them in the wastebasket. Cost?—sixty cents a dozen, a nickel apiece, and laundering costs seven cents per collar!

As for new fabrics—the future belongs to specialized, made-to-order materials. Cotton men pin their faith on this cheap, strong, washable, versatile fiber to be the base upon which the new combination will be built.

Nylon is now supplemented with plastics, Saran and Vinylite, in filaments and narrow strips. Metal thread, aluminum and zinc, are being used. Crinkles and crimps and crepes are being permanently woven into goods for novel effects. The combinations and permutations are infinite. With growing knowledge of the complicated chemistry of fibers, plastics, and synthetic rubbers, it is going to be possible to combine the molecules to give to cloth entirely new properties. Southern textile research is dipping eager fingers into a regular Jack Horner pie and is pulling out some luscious plums.

The epidemic of research that is sweeping the country and which has so virulently attacked the Southern textile men disturbs some of the best friends of applied science. They mistrust our American fondness for fads and are fearful of a new type of racket. They snort scornfully at the "research departments" set up by advertising agencies, organized charities, and a myriad of Government bureaus. They are disdainful of research over the air and in the annual report of many a corporation. To count the diaper pins sold in the dime stores of Seattle and call it "research," they point out, is stretching a good word into a catch phrase. Averaging pounds of Swiss cheese per person, correlating the popularity of cocker spaniels and boxers, and all such interesting, sometimes enlightening fact-findings are not research at all, and the scientist is quite right to protest this misappropriation of a term that implies not only the discovery of facts, but their confirmation by objective, controlled experiment.

Nevertheless, this lively, popular interest in research is healthy, encouraging, and long overdue, and its sudden outburst in the textile industry is not exactly typical. However willing the spirit, the flesh of most Southern industries is weak. But this is typical of almost the whole nation.

The lack of true research and the need for it in the South was the text from which Dr. George D. Palmer, Jr., preached to the assembled manufacturers at the 1940 meeting of the Alabama State Chamber of Commerce. It was a good, workmanlike job, well turned out. The professor of chemistry at the University of Alabama had the facts and presented them clearly. But he is no soul-stirring orator and he was telling an oft-told tale. This time, however, the familiar theme struck fire; Palmer's sermon made converts who burned with fiery zeal.

The embers had been smoldering a dozen years. When Herbert Ryding was president of the Tennessee Coal, Iron & Railroad Company, he had lighted the match by asking Dr. Stuart J. Lloyd, dean of the School of Chemistry, University of Alabama, to study the chemical potentialities of the Birmingham section. That lean chemical zealot urged, and kept urging, that businessmen of the South band together to support an industrial research organization upon the lines of the

Mellon Institute. Something almost happened in Mobile, April 1940, at the Southern Association of Science and Industry meeting. Six months later definite action was taken by the Alabama Chamber of Commerce. A committee was appointed.

Such committees do not always function, but this one had the right chairman. Thomas W. Martin had the vision of what research will do for Southern industry and the energy to incarnate this idea. Native Alabamian, successful attorney, for twenty years president of the Alabama Power Company, Tom Martin knew by his first name practically every likely citizen who might back such a project. Alabamians like him and respect him and they turned over to him nearly a million in cash and pledges, the first three years' support of this research idea.

The Alabama Research Institute quickly grew into the Southern Research Institute. Wallace L. Caldwell headed a committee that winnowed a hundred and four possible candidates for the key post of director, to select Dr. Wilbur A. Lazier. A launching party was held in Birmingham, October 4, 1944; a dinner at which Dr. Edward R. Weidlein, director of the Mellon Institute, who had generously been father confessor in maturing these ambitious plans, made an inspiring address, and the new director was introduced.

To that distinguished gathering, Tom Martin made clear the need of research in the South in cold figures:

65 per cent of the nation's manufactured wares are produced in the territory east of the Mississippi and north of the Ohio, where 78.2 per cent of the patents are issued and 89.2 per cent of the research personnel work—against 8.7 per cent of the manufactured production, 2.9 per cent of the patents, and 2.2 per cent of the research workers in the Southeastern states. Fewer patents and only a few more researchers have been in these nine states than in Connecticut alone, which stands fifth on the national roll.

After sketching graphically the opportunities of cooperative research, he concluded: "We should feel some satisfaction that we who represent private enterprise have willingly joined in this effort under the management of men of business whose sole return must be the conscious feeling that they are doing something for the region, that they are doing it in the spirit of enterprise and initiative, recognizing that as the region prospers, so will all business and industry."

That justly proud statement is more than an expression of the vivifying conception of self-help for the South. It comprehends a fundamental truth in applied research. American industry dare not leave research to the Government.

This plain fact has been dangerously camouflaged. Quite aside from any qualms over control or regimentation, ignoring the "dead hand of bureaucracy" which is so throttling to the spirit of scientific inquiry, Government research cannot possibly achieve workaday results that the American people must have if we are to advance our material status.

Inherent in public research are conditions that in a democracy sterilize its efforts. No matter how competent the staff, how elaborate the laboratory, how ample the funds, practical results must always be pitiful. Two insurmountable obstacles block the way to progress.

Pioneering research is a gamble. If one project in a hundred hits the commercial target, the marksmanship is in the sharpshooter class. Even when an industrially feasible discovery is made, it takes from five to eight years to bring it from test tube to sales counter. Risk and time are so inevitably prerequisites that the sharp phrase "patient money" has been coined to describe the venture capital needed to finance research programs. No democratic government has this kind of funds available. Tax money is not patient money.

Moreover, every discovery of the laboratory threatens some vested interest. A million and a half is a picayune in this year's federal budget. But what Congress would vote even this trifling sum for eight consecutive years to create a competitor for our established rayon factories and our distressed cotton plantations? Yet in years and dollars that was the initial cost of the development of nylon. The answer is ludicrously self-evident.

Just to clinch the point, rayon cord is demonstrably better and safer than cotton duck in heavy-duty tires for trucks and planes, yet in the midst of war the Southern wing of the farm bloc crusaded for a law compelling the Army to use the inferior material. The farm hullabaloo in Washington to continue the production of butadiene from grain alcohol for synthetic rubber—a useful wartime expediency, but a more costly process than from petroleum—proves the same point.

If we entrust to our Government the national research that is to develop new products and processes and create new industries, we court two disasters. We hand over the key to our future prosperity to the politicians. We subject our elected representatives to enormous pressures from many different minority interests. The first would be terrifically costly; the second, extremely dangerous. Nothing in the recent record either of the wartime governmental agencies or of the Congressional blocs prompts us to encourage either of these too-plain tendencies.

The Southern Research Institute is a natural and effective counter to the insidious proposal, so rabidly supported by Senator Kilgore and all the Washington bureaucrats, that the Government unify and direct all research for the common good. The Institute encourages cooperation, but it maintains individual initiative. Since it is not a money-making organization, its staff and scientific equipment offer effective means of research at a cost within the reach of the smaller manufacturers who can rarely afford to maintain either, yet who so often most sorely need this assistance.

The record of the first year is enheartening: the initial laboratories enviably well equipped and manned by a staff of twenty highly trained, competent research workers. Half of these are Southerners, born, bred, and educated—a retort courteous to the complaint that the best technical brains of the South have been enticed away to more tempting opportunities in the North and West. Six months after it opened its doors for business, the business of research for Southern industries, the Institute reports twenty sponsors and project contracts in excess of $200,000. Cotton textiles, cottonseed products, tobacco, peanuts, citrus by-products, essential oils, metallurgy, and mechanical problems are all on the laboratory work benches. Already, men and apparatus jostle each other, and enthusiastic friends are bustling out to raise $2,500,000 to build and equip a seven-story laboratory where two hundred scientists can hunt answers for Southern sponsors.

The *modus operandi* is simple and direct. The sponsor places a definite project with the Institute to be carried forward at an agreed rate. A qualified scientist with all the facilities of the Institute behind him is assigned to the project, and expenses beyond the agreed sum may not be incurred without the sponsor's approval. The field of the research is defined and any and all discoveries or inventions within its scope, as well as all information developed, are promptly made known to the sponsor and become his exclusive property. Unless released by the sponsor, all the work is confidential, carried on under the inviolate seal of professional ethics. To the sponsor these last two points are vital, but they are important, too, in the commercial development of any discoveries made.

Workers in government laboratories are bound to take out "public interest patents" for their discoveries. The Government then offers "shop rights" to any interested manufacturer. Theoretically, this is the perfect procedure for the common good. Practically, it often delays the commercial exploitation of valuable patents since the interest displayed by manufacturers in public patents is cooler than tepid.

As an illustration, Professor Schoch of the University of Texas has patented a process that reaches out boldly toward the chemical application of atomic energy, a sort of little second cousin of the atomic bomb. By bombarding natural gas with electrical discharges he produces acetylene. His discovery is not only interesting in itself, but it opens up entirely new methods of inducing chemical reactions. However, this significant patent is so involved with the rights of both the University and the state of Texas that experienced experts at the highest executive level of the three companies which are most logical exponents of this invention have confessed that the legal and political entanglements have strangled their interest in this development. Because of lack of adequate patent protection half a dozen likely new products and some potential new processes developed at the various Regional Laboratories of the Department of Agriculture have been peddled without finding a strong commercial sponsor. This dead-end result is not exactly "in the public interest."

Advocates of the Kilgore Bill, who insist so pointedly on the public ownership of all medical discoveries, should regard thoughtfully the plight of pharmaceutical-chemical research in France. This seemingly benevolent idea is embodied in the French patent law which does not protect medical discoveries. As a result this branch of chemistry has withered. When by chance some French chemist stumbles upon a promising therapeutic agent, he smuggles the formula to Germany or England where he can patent his discovery.

On top of the risks and the costs of research must be piled the risks and costs of bringing up the laboratory child to commercial maturity in a manufacturing plant. Pyramided above both are the risks and costs of market development. The long gamble of time and money must be safeguarded or nobody will take all these chances.

To the South, especially, the future of that pioneering research which creates new industries is today crucial. Broader than this immediate sectional consideration, research is the key to the continued economic progress of the whole country. This scientific tool is the best means man has ever found for providing permanent, purposeful employment in the production of usable goods and for building up a surplus out of which to yield a more abundant life. Research is thus the only valid guarantee of the future, the underwriter that will make good the promises of the new spirit that inspires Southerners, the broad highway to the brighter horizons for all Americans.

Acknowledged with Thanks

TO THE MEN whose words and deeds are the very warp and woof of this book my gratitude is as great as my obligations are evident. Without them the foregoing pages could never have been written. To them I can say no more, but there are others whose names do not appear in the Index of this book who have helped me. Some paved my way to strangers with introductions, others supplied me with facts and figures, still others have read parts of the manuscript critically and given me the benefit of their corrections and suggestions. To all these, my sincere thanks:

W. W. Ball
Melchior Beltzhoover
Ralph W. Bost
Harry E. Brants
Alexander Calder
T. A. Cook
W. B. Cotton
Douglas Coutler
Russell H. Dunham
E. K. Earkert
E. B. Erard
Robert C. Goodwin
A. A. Green, Jr.
W. F. Guinee
Claudia L. Hamm
J. Ross Hanahan
S. M. Harmon
Franklin R. Hoadley
Giles E. Hopkins
W. M. Hutchinson
Floyd Jefferson
Russell W. Kerm

Harold A. Levey
Ernest L. Little
R. G. MacDonald
Wheeler McMillen
Theodore Marvin
Claudius Murchison
Robert C. Palmer
Oliver Porter
Harley W. Ross
J. P. Sheehy
Floyd Studer
T. J. Twomey
Vincent F. Waters
Otto O. Watts
P. F. Watzel
Hendricks Whitman

A host of unwitting collaborators have also made their contributions: storekeepers and farmers, gas station attendants and hotel bellboys, waitresses and receptionists, people I met on trains, in plants and laboratories, at lunch and dinner.

I am notably indebted to Ben Hibbs, Editor of *The Saturday Evening Post,* whose commission sent me through the South and by whose permission material from articles appears in three chapters.

WILLIAMS HAYNES

Index

INDEX

INDEX

INDEX

INDEX

INDEX

INDEX

INDEX

INDEX

SOUTHERN RESOURCES

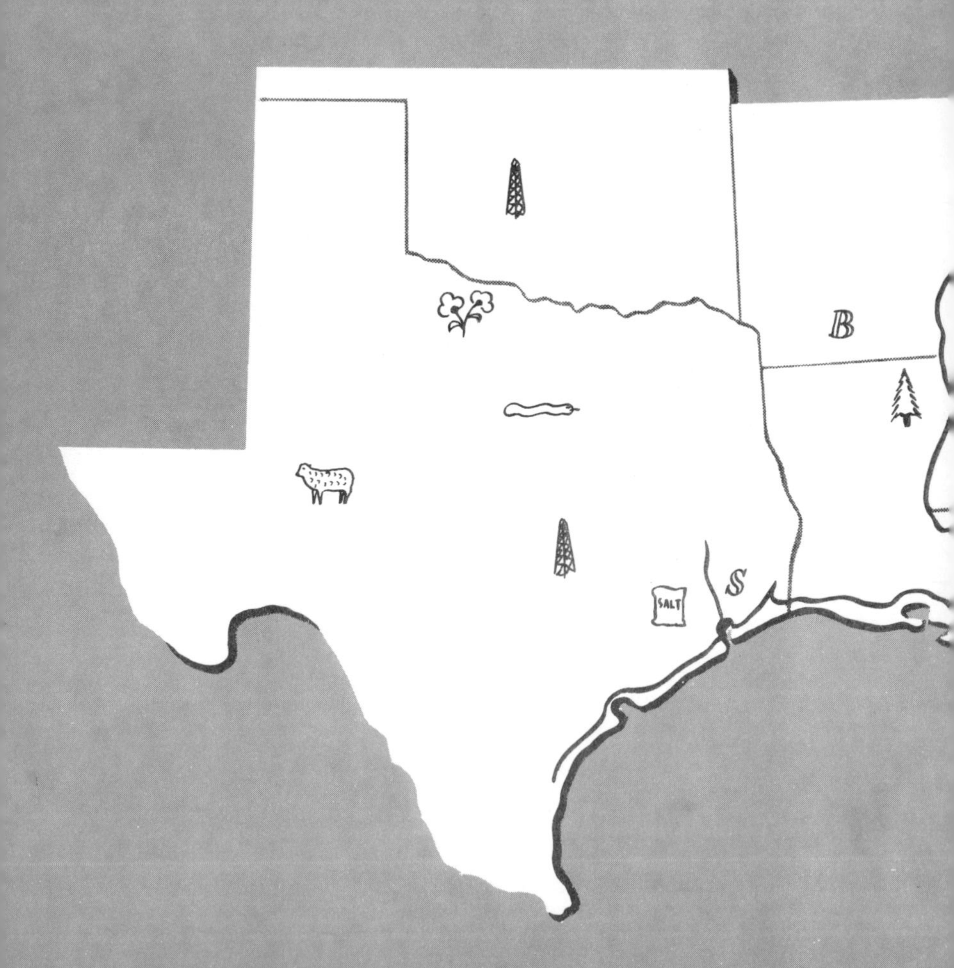

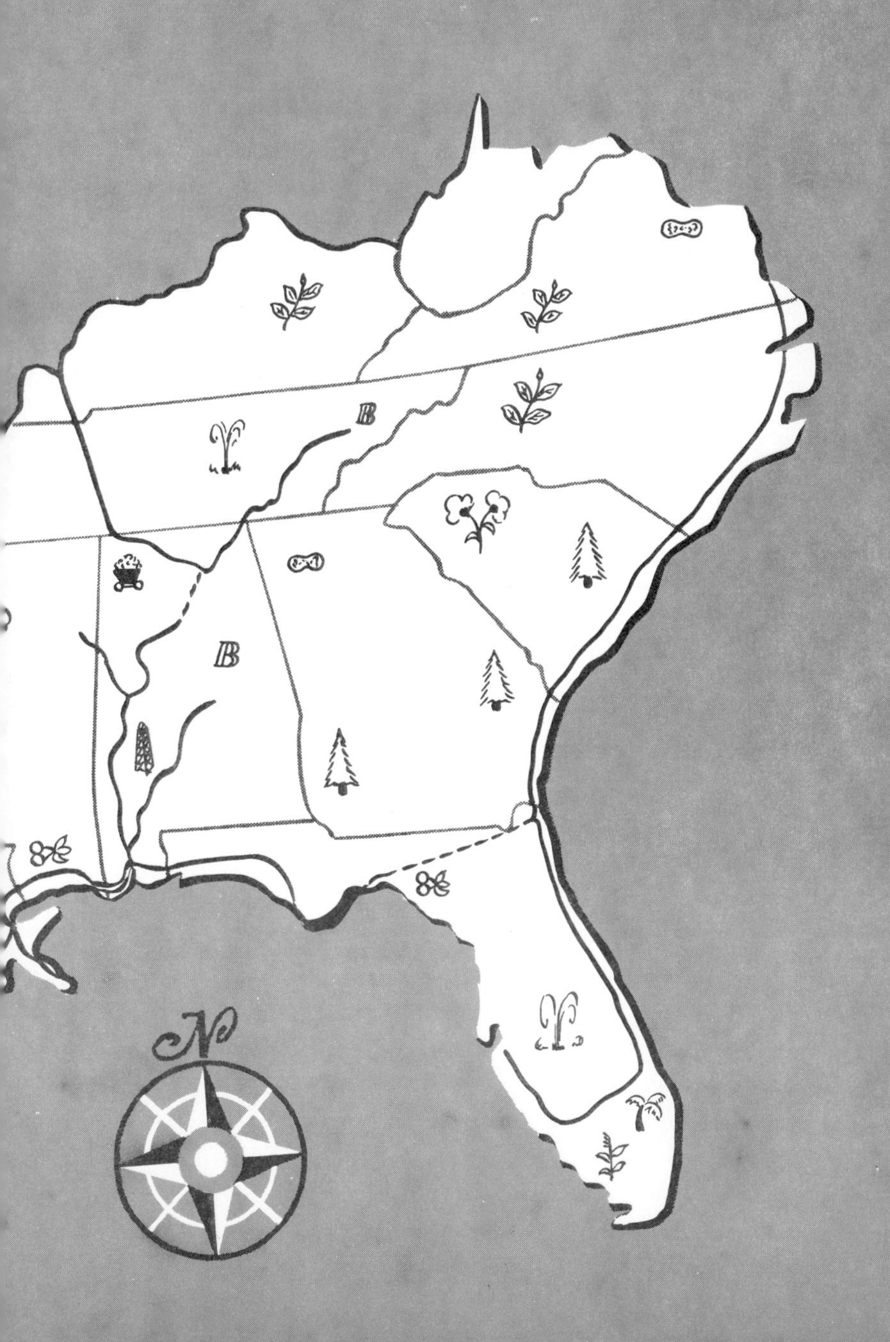
N